# Fundamentals of DNA Methylation And Basic Mechanisms

# Fundamentals of DNA Methylation And Basic Mechanisms

Kamal Duggal

CYBER TECH PUBLICATIONS
4264/3, Ansari Road, Daryaganj, New Delhi-110002 (India)
Ph.: 011-23244078, 011-43559448 Fax: 011-23280028
E-Mail: cyberpublicationsdelhi@yahoo.com
Website: www.cybertechpublications.com

Fundamentals of DNA Methylation And Basic Mechanisms

**Kamal Duggal**



First Edition 2011



*Published by :*

**G.S. Rawat** for **Cyber Tech Publications**
4264/3, Ansari Road, Daryaganj, New Delhi-110002 (India)
Ph.: 011-23244078, 011-43559448 Fax: 011-23280028
E-Mail: cyberpublicationsdelhi@yahoo.com
Website: www.cybertechpublications.com

# Contents

# Preface

"Microbiology is an exceptionally broad discipline encompassing specialties as diverse as biochemistry, cell biology, genetics, taxonomy, pathogenic bacteriology, food and industrial microbiology, and ecology. A microbiologist must be acquainted with many biological disciplines and with all major groups of microorganisms: viruses, bacteria, fungi, algae, and protozoa. The key is balance. Students new to the subject need an introduction to the whole before concentrating on those parts of greatest interest to them. This text provides a balanced introduction to all major areas of microbiology for a variety of students. Because of this balance, the book is suitable for courses with orientations ranging from basic microbiology to medical and applied microbiology. Students preparing for careers in medicine, dentistry, nursing, and allied health professions will find the text just as useful as those aiming for careers in research, teaching, and industry.

This text is designed to be an effective teaching tool. A text is only as easy for a student to use as it is easy to read. Readability has been enhanced by using a relatively simple, direct writing style, many section headings, and an organized outline format within each chapter. The level of difficulty has been carefully set with the target audience in mind. During preparation of the fifth edition, every sentence was carefully checked for clarity and revised when necessary. The American Society for Microbiology's ASM Style Manual conventions for nomenclature and abbreviations have been followed as consistently as possible.

The many new terms encountered in studying microbiology are a major stumbling block for students. This text lessens the problem by addressing and reinforcing a student's vocabulary development in three ways: (1) no new term is used without being clearly defined (often derivations also are given)--a student does not have to be familiar with the terminology of microbiology to use this text; (2) the most important terms are printed in boldface when first used; and (3) a very extensive, up-to-date, page-referenced glossary is included at the end of the text.

Because illustrations are critical to a student's learning and enjoyment of microbiology, all illustrations are full-color, and many excellent color photographs have been used. Color not only enhances the text's attractiveness but also increases each figure's teaching effectiveness. Considerable effort has gone into making the art as attractive and useful as possible. Much of the art in the fourth edition has been revised and improved for use in the fifth edition. All new line art has been produced under the direct supervision of an art editor and the authors, and designed to illustrate and reinforce specific points in the text. Consequently every illustration is directly related to the narrative and specifically cited where appropriate. Great care has been taken to position illustrations as close as possible to the places where they are cited. Illustrations and captions have been reviewed for accuracy and clarity.

— *Author*

# 1

# Introduction

## DEVELOPMENT, GENETIC DISEASE AND CANCER – DEVELOPMENT – GENETIC DISEASE – CANCER

The second volume on 'DNA Methylation: Basic Mechanism' in the series Current Topics in Microbiology and Immunology will follow in 2006.

In assembling these chapters and editing the two volumes, we intend to address the rapidly growing number of—particularly young—researchers with an interest in many different areas of biomedicine. Particularly, for our colleagues in molecular medicine, a sound basic knowledge in the biology and biochemistry of DNA methylation will prove helpful in critically evaluating and interpreting the functional meaning of their findings in medical genetics and epigenetics or in cancer research. The authors of the current chapters invariably point to the complexity ofproblems relatedto DNA methylation and our still limited understanding of its function. A healthy caveat will therefore be in order in the interpretation of data related to medical problems.

The structural and functional importance of the "correct" patterns of DNA methylation in all parts of a mammalian genome is, unfortunately, not well understood. The stability, inheritability, and developmental flexibilityof these patterns all point to a major role that these patterns appear to play in determining structure and function of the genome. Up to the present time, studies on the repetitive sequences, which comprise >90% of the DNA sequences in the human or other genomes, have been neglected. We only have a vague idea about the patterns of DNA methylation in these abundant sequences, except that the repeat sequences are often hypermethylated, and that their patterns are particularly sensitive to alterations upon the insertion of foreign DNA into an established genome.

Upon foreign DNA insertion into an established genome, during the early stages of development, or when the regular pathways of embryonal and/or fetal development are bypassed, e.g., in therapeutic or reproductive cloning, patterns ofDNA methylation in vast realms ofthe genome can be substantially altered. There is very little information about the mechanisms and conditions of these alterations, and investigations into these areas could be highly informative. By the same token, a thorough understanding of these problems will be paramount and a precondition to fully grasp the plasticity of mammalian genomes. Moreover, it is hard to imagine that, without this vital information at hand, we will be successful in applying our knowledge in molecular genetics to the solution of medical problems. A vast amount of basic research still lies ahead of us. I suspect that, in the hope of making"quick discoveries" and, consequently, in neglecting to shoulder our basic homework now, we will only delay the breakthroughs that many among us hope for.

## On the Early History of 5-mC

In thefall of 1966, Norton D. Zinder of Rockefeller University in New York City presented the Harvey Lecture on "Phage RNA as Genetic Material" (Zinder 1966). Frankly, I do not remember many details of his talk. However, one of his concluding remarks, in which he thanked his teacher Rollin Hotchkiss, stuck in my mind and became an important leitmotif for much of my own scientific career. Norton's relevant passages went something like this (approximate quotation):

When we hope to have made a scientificdiscovery, we better spendmuch of our time immediately after this fortunate event in trying to counter our own beliefs and interpretations. Only after a lot of painstaking scrutiny involving many control experiments when our discovery has stood the test of careful consideration, can one hope that our colleagues willbeabletoconfirmthe newfindings. Ofcourse, it isa major taskofthe scientific community to respectfully meet supposedly novel announcements with disbelief and skepticism and in turn commence the process of disproving these concepts. Consistent confirmations, with plenty of modifications to be sure, will provide the encouragement necessary to continue and to improve the initial observations and conclusions.

Apparently, the scientific tradition reflected in this overall cautious attitude had emanated from the laboratory of Oswald Av e r y t hat Rollin Hotchkiss had been trained in. This certainly most important of scientific credos seems to contradict intuitively held notions and might be thought to run counter to general practice. Today, Avery's p hilosophy towards scientific research sometimes seems ages remote from the fast-hit mentality of the "impact factor" generation. And yet, one had better heed his advice.

Long-standing experience with the early, and for this matter present, studies on the biological function of DNA methylation in eukaryotic systems constitutes a case in point. Many observations, although recorded correctly, had tobe frequently re-interpreted. The generality of the functional importance of the fifth nucleotide was often questioned, frequently by researchers working on *Drosophila melanogaster* who only recently learned that during embryonic development of this organism, 5-mC also makes an appearance (Lyko et al. 2000). Even initially sound skepticism has sometimes to be re-evaluated.

The fifth nucleotide, 5-methyl-deoxycytidine (5-mC), was first described in DNA from the tubercle bacillus (Johnson and Coghill 1925) and in calf thymus DNA (Hotchkiss 1948). I cite from the article by Rollin Hotchkiss, 1948, in the *Journal of Biological Chemistry*:

> This small peak has been observed repeatedly in the chromatographic patterns from acid hydrolysates of a preparation of calf thymus deoxyribonucleic acid.... In this connection it might be pointed out that 5-methyl cytosine was reported by Johnson and Caghill as a constituent of the deoxyribonucleic acid of the tubercle bacillus.

Subsequently, 5-mC had a biochemical future as 5-hydroxymethyl-C (5-hm-C) in the DNA of the T-even bacteriophages. The biological function of this C modification was never elucidated. Daisy Dussoix and Werner Arber (Arber and Dussoix 1962; Dussoix and Arber 1962) discovered the phenomena of restriction and modification in bacteria. It was recognized later that DNA modifications, like 5-mC and/or $N^6$-methyl adenosine ($N^6$-mA), had important biological consequences. A major endeavor followed in many laboratories that worked on the biochemistry of DNA modifications in bacteria and their phages (review by Arber and Linn 1969). Around 1970, Hamilton Smith and his colleagues discovered the restriction endonucleases (Kelly and Smith 1970) whose application to the analyses of DNA was pioneered by Daniel Nathan's laboratory (Danna and Nathans 1971). It was soon appreciated that enzymes, whose activity was compromised by the presence of a 5-mC or an $N^6$-mA in the recognition sequence, could be of great value in assessing the methylation status of a DNA sequence.

In their investigations on the globin locus, Waalwijk and Flavell (1978) have observed that the isoschizomeric restriction endonuclease pair HpaII and MspI both recognize the sequence 5'-CCGG-3', and hence can be used to test for the presence of a 5-mC in this sequence. HpaII does not cleave the methylated sequence, whereas MspI is not affected in its activity by methylation. To this day, cleavage by this enzyme pair provides a first approach to the analysis of methylation patterns in any DNA. A useful review (Mc-Clelland and Nelson 1988) summarizes the specificities of a large number of methylation-sensitive restriction endonucleases.

In 1975, two papers (Holliday and Pugh 1975; Riggs 1975) alerted the scientific community to the importance of methylated DNA sequences in eu-karyotic biology. Our laboratory at about that time, independently, analyzed DNA in the human adenovirus and in adenovirus-induced tumor cells for the presence of 5-mC residues (Günther et al. 1976) and discovered that integrated adenovirus DNA—perhaps any foreign DNA—had become de novo methylated (Sutter et al. 1978). DNA methyltransferases in human lymphocytes were studied early on by Drahovsky and colleagues (1976). Vanyushin's (1968) laboratory in Moscow analyzed the DNA of many organisms for the presence of 5-mC and $N^6$-mA.

It soon became apparent that bythe use of methylation-sensitive restriction endonucleases only a subset of all 5'-CG-3' dinucleotides would be amenable to methylation analysis. Depending on the nucleotide sequence under investigation, only 10% to 15%—or even fewer—of these dinucleotide sequences could be screened for methylation by the combined application of HpaII/MspI and HhaI (5'-GCGC-3'). Church and Gilbert (1983) were the first to develop a genomic sequencing technique, based on the chemical modification of DNA by hydrazine, and thus provided a means to survey all possible C-residues for the occurrence of 5-mC in a sequence. The bisulfite sequencing technique introduced by Marianne Frommer and colleagues (Frommer et al. 1992; Clark et al. 1994) allowed for a positive display of methylated sequences. This method, along with some of its modifications, has now become the "gold standard" in analytical work on DNA methylation. The method is precise and yields reproducible results but is laborious and expensive. At the moment, however, there is no better method available.

Constantinides, Jones and Gevers (1977) reported that the treatment of chicken embryo fibroblasts with 5-aza-cytidine, a derivative of cytidine that was known to inhibit DNA methyltransferases (review by Jones 1985), activated the developmental program in these fibroblasts leading to the appearance of twitching myocardiocytes, adipocytes, chondrocytes, etc. in the culture dish. Their interpretation, at the time, that alterations in DNA methylation patterns activated whole sets of genes involved in realizing a developmental program, has stood the test of time. There is now a huge literature on changes in DNA methylation during embryonal and fetal development (for an early contribution to this topic, see Razin et al. 1984).

The observation on inverse correlations between the extent of DNA methylation and the activity of integrated adenovirus genes in adenovirus type 12-transformed hamster cells (Sutter and Doerfler 1980a, b) elicited a surge of similar investigations on a large number of eukaryotic genes. Today, it is generally accepted that specific promoter methylations in conjunction with histone modifications (acetylation, methylation, etc.) play a crucial role in the long-term silencing of eukaryotic genes (Doerfler 1983). There is no rule, however, without exceptions: Willis and Granoff (1980) have shown that the genes of the iridovirus frog virus 3 (FV3) are fully active notwithstanding the complete 5'-CG-3' methylation of the virion DNA and of the intracellular forms of this interesting viral genome.

Since many foreign genomes in many biological systems and hosts frequently became de novo methylated, several authors have speculated whether this phenomenon reflects the function of an ancient cellular defense mechanism against the uptake and expression of foreign genes (Doerfler 1991; Yoder et al. 1997) much as the bacterial cell has developed the modification restriction systems to counter the function of invading viral genomes. In eukaryotes, integrated foreign viral, in particular but not exclusively, retro-transposon genomes, which make up a huge proportion of the mammalian and other genomes, are frequently hypermethylated. This finding obviously is in keeping with the cellular defense hypothesis of de novo methylation mechanisms. In our laboratory, these considerations have prompted investigations on the stability of food-ingested DNA in mammals as a possible source of foreign DNA taken up with high frequency by mammalian organisms (Schubbert et al. 1997; Forsman et al. 2003).

How have the patterns of DNA methylation, that is the distribution of 5-mC residues in any genome, evolved over time? How different are these patterns from cell type to cell type and under what conditions are they preserved, even interindividually maintained, in a given species? In what way do these patterns co-determine the structure of chromatin by providing a first-line target for proteins binding preferentially to methylated sequences (Huang et al. 1984; Meehan et al. 1989) or by being repulsive to specific protein-DNA interactions?

Chromatin structure and specific patterns of DNA methylation, which differ distinctly from genome region to genome region, are somehow related. There is growing experimental evidence that the presence of 5-mC residues affects the presence of a large number of proteins in chromatin. However, we do not understand the actual complexity of these interactions or the role that histone modifications can play in conjunction with DNA methylation in the control of promoter activity. Imaginative speculations abound in the literature, but there is little novel experimental evidence. I suspect we will have to unravel the exact structural and functional biochemistryof chromatin before real progress on these crucial questions will become possible. A recent review (Craig 2005) phrases the chromatin enigma thus "... there are many different architectural plans..., leading to a seemingly never-ending variety of heterochromatic loci, with each built according to a general rule."

With the realization and under the premise that promoter methylation could contribute to the long-term silencing of eukaryotic genes, researchers approached the fascinating problem of genetic imprinting. Several groups at that time provided evidence that genetically imprinted regions of the genome can exhibit different methylation patterns on the two chromosomal alleles (Sapienza 1995; Chaillet et al. 1995). For one of the microdeletion syndromes involving human chromosome 15q11-13, Prader-Labhart-Willi syndrome, a molecular test was devised on the basis of methylation differences between the maternallyand the paternallyinherited chromosome (Dittrich et al. 1992).

Problems of DNA methylation, of the stability and flexibility of the patterns of DNA methylation are also tightly linked to many unresolved questions of reproductive and/or therapeutic cloning. In an effort to correlate gene expression with survival and fetal overgrowth, imprinted gene expression in mice cloned by nuclear transfer or in embryonic stem (ES) cell donor populations from which they were derived has been investigated. The epigenetic state of the ES cell genome appears to be extremely unstable. Variation in imprinted gene expression has been observed in most cloned mice. Many of the animals survived to adulthood despite widespread gene dysregulation, indicating that mammalian development may be rather tolerant to epigenetic aberrations of the genome. These data imply that even apparently normal cloned animals may have subtle abnormalities in gene expression (Humpherys et al. 2001). In cloned animals, lethality occurs only beyond a threshold of faulty gene reprogramming of multiple loci (Rideout et al. 2001). Of course, malformations are frequent among cloned animals, which appear to have also a limited lifespan.

Similarly, the idea to replace defective genes with their wild-type versions or to block neoplastic growth by introducing cogently chosen genes and stimulate the defenses against tumors and metastases has captured the fascination of many scientists working towards realistic regimens in gene therapy. However, many unsolved problems have remained with viral gene transfer vectors:

(1) Stable DNA transfer into mammalian cells was frequently inefficient.

(2) The site of foreign DNA insertion into the recipient genomes could not be controlled.

(3) The integrates at random sites were often turned off unpredictably due to cellular chromatin modifications and/or the de novo methylation of the foreign DNA.

Of course, there had been prominent voices cautioning against the premature application of insufficiently scrutinized concepts and techniques (cited in Stone 1995). Adenovirus vectors proved highly toxic in topical applications to the bronchial system of cystic fibrosis patients (Crystal et al. 1994). In a tragic accident, the administration of a very high dose of a recombinant adenovirus, which carried the gene for ornithine-transcarbamylase, led to the death of 18-year-old Jesse Gelsinger. Retroviral vectors as apparent experts in random integration were thought to assure continuous foreign-gene transcription in the target cells. By using a retroviral vector system, 10 infant boys suffering from X-linked severe combined immunodeficiency (X-SCID) had presumably been cured. However, the scientific community was alarmed soon thereafter by reports that 2 of these infants developed a rare T cell leukemialike condition (Hacein-Bey-Abina et al. 2003). Presumably, the integration of the foreign DNA construct had activated a protooncogene in the manipulated cells—perhaps a plausible explanation and in line with long-favored models in tumor biology.

The possibility exists that the insertion of foreign DNA into established mammalian genomes, with a preference at actively transcribed loci, can alter the chromatin configuration even at sites remote from those immediately targeted by foreign DNA insertion (Doerfler 1995, 2000). In cells transgenic for adenovirus or bacteriophage lambda DNA, extensive changes in cellular DNA methylation (Heller et al. 1995; Remus et al. 1999) and cellular gene transcription patterns (Müller et al. 2001) have been documented. Foreign DNA insertion at one site may, hence, affect the genetic activity of a combination of loci that can be disseminated over the entire genome. The chromosomal sites of the cellular genes thus af-flictedmightdependon thelocation ofthe initialintegration event. Oncogenic transformation of the cell, according to this model, would ensue because of alterations in specific combinations ofgenes and loci and in extensive changes in the transcriptional program of many different genes.

If valid, this concept could shed doubts on apparently useful procedures in molecular medicine—the generation of transgenic organisms, current gene therapy regimens, perhaps even on the interpretation of some knock-out experiments. The functional complexities of the human, or any other, genome cannot yet be fathomed by the knowledge of nucleotide sequences and the current textbook wisdom of molecular biology. At this stage of our "advanced ignorance" in biology, much more basic research will be the order of this and, I suspect, many future days in order to be able to heed the primary obligation in medicine—*nil nocere.*

## ONWARD TO NEW PROJECTS

By now, the concept of an important genetic function for 5-mC in DNA has been generally accepted. Moreover, many fields in molecular genetics have included studies on the fifth nucleotide in their repertoire of current research: regulation of gene expression, structure of chromatin, genetic imprinting, developmental biology (even in *Drosophila melanogaster*, an organism whose DNA hasbeenpreviously thoughttobedevoidof5-mC),cloningoforganisms, human medical genetics, cancer biology, defense strategies against foreign DNA, and others. Progress in research on many of these topics has been rapid, and the publication of a number of concise reports within the framework of *Current Topics* is undoubtedly timely. When screened for "DNA methylation" in October 2005, PubMed responds with a total of 9,772 entries dating back to 1965; a search for "DNA methylation and gene expression" produces 4,167 citations.

A conventional review article on DNA methylation or on one of its main subtopics, therefore, would have to cope with serious limitations, omissions and over-simplifications. With more than 30 years of experience in active research in the field, I wish to briefly outline questions, problems, and possible approaches for further research. Seasoned investigators in the field undoubtedlywill have their own predilections. For the numerous newcomers to studies on DNA methylation, my listing might provide an introduction or more likely might arouse opposition that will be just as useful in aiding initiate original research.

### *Chromatin Structure*

Patterns of DNA methylation in the genome and the topology of chromatin structure and composition are tightly linked. Studies on the biochemical modifications of histones—amino acid sequence-specific acetylations and methylations (Allfrey et al. 1964; and more than 3,100 references afterwards) have revealed the tip of the iceberg. A much more profound understanding of the biochemistry of all the components of chromatin and their possible interactions with unmethylated or methylated DNA sequences will have to be elaborated. I would rate such studies as the No. 1 priority and primary precondition for further progress in the understanding of the biological significance of DNA methylation.

### *Promoter Studies*

We still donot understand the detailsofhow specific distributions of 5-mC residues in promoter or other upstream and/or downstream regulatory sequences affect promoter activity. It is likely, though still unproved, that there is a specific pattern for each promoter, perhaps encompassing only a few 5'-CG-3' dinucleotides, that leads to promoter inactivation.

It would be feasible to modify one of the well-studied promoters in single or in combinations of 5'-CG-3' sequences and follow the consequences for promoter activity with an indicator gene. Moreover, for each methylated 5'-CG-3' sequence, the promotion or inhibition of the binding of specific proteins, transcription factors, and others will have to be determined.

It is still unpredictable whether there is a unifying system applying to classes of promoters or whether each promoter is unique in requiring specific combinations of 5'-5m-CG-3' residues for activity or the state of inactivity. Of course, in this context, the question can be answered of whether the activity of a promoter can be ratcheted down by methylating an increasing number of 5'-CG-3' dinucleotides step by step in increments of one.

## *Correlations between DNA Methylation and Histone Modification in Eukaryotic Promoters*

In what functional and enzymatic ways are these two types of modifications interrelated? Can one be functional without the other? Is one the precondition for the other one to occur? Ever since the search began for the class of molecules that encodes genetic information, the "battle has raged," as it were, between proteins and DNA to exert the decisive impact.

A similar, though much less fundamental, debate on the essential mechanisms operative in long-term gene inactivation is occupying our minds today. In most instances, the 5-mC signal is relevant mainly in long-term gene silencing. For frequent fluctuations between the different activity states of a promoter, the DNA methylation signal would be a poor candidate for a regulatory mechanism, because promoter methylation is not easily reversible.

## *On the Mechanism of de Novo Methylation of Integrated Foreign or Altered Endogenous DNA*

One of the more frequent encounters for molecular biologists with DNA methylation derives from the analysis of foreign DNA that has been chro-mosomally integrated into an established eukaryotic genome.

Foreign DNA can become fixed in the host genome not only after the infection with viruses but also in the wake of implementing this integration strategy in the generation of transgenic organisms. In knock-in and knock-out experiments, in regimens of gene therapy, and others, investigations on this apparently fundamental cellular defense mechanism against the activity of foreign genes—de novo methylation—has both theoretical and practical appeal. During the embryonic development of mammals, methylation patterns present at very early stages are erased and new patterns are reestablished de novo in later stages. Hence, we lack essential information on a very important biochemical mechanism. There are only a few systematic studies on the factors that influence the generation of de novo methylation patterns. Size and nucleotide sequence of the foreign DNA as well as the site of foreign DNA insertion could have an impact, but in what way remains uncertain. Other aspects of de novo methylation relate to the availability, specificity, and topology of the DNA methyltransferases in the chromatin structure.

## *Levels of DNA Methylation in Repetitive DNA Sequences*

Studies on repetitive DNA sequences and their functions constitute one of the very difficult areas in molecular biology, mainly for the want of new ideas to contribute to the investigations. Perhaps the elucidation of the patterns of 5-mC distribution in these sequences could shed light on possible novel approaches of how to proceed further. Repetitive DNA sequences, particularlyretrotransposon-derived DNA or endogenous retro-viral sequences, are in general heavily methylated. Exact studies on the methylation and activity of specific segments in the repetitive DNA are available only to a limited extent. The difficulty for a systematic analysis certainly lies in the high copy number and the hard to disprove possibility that individual members of a family of repetitive sequences might exhibit different patterns.

## *Foreign DNA Insertions can Lead to Alterations of DNA Methylation in* Trans

Studies on this phenomenon have occupied our laboratory for several years, and we are still investigating whether these alterations might be a general consequence of foreign DNA insertions or occurred only under distinct conditions. We, therefore, propose to pursue the following strategies.

(a) Random insertion of a defined cellular DNA segment with a unique orarepetitive sequenceatdifferent chromosomalsites andfollow-up of changes in DNA methylation in different locations of the cellular genome. In this context, methylation patterns in unique genes andin retrotransposons or other repetitive sequences will be determined.

(b) In individual transgenic cell clones, transgene location should be correlated with methylation and transcription patterns in the selected DNA segments. Could the chromosomal insertion site of the transgene be in contact with the regions with altered DNA methyla-tion on interphase chromosomes?

(c) Studies on histone modifications in or close to the selected DNA segments in which alterations of DNA methylation have been observed.

(d) Influence of the number of transgene molecules, i.e., the size of the transgenic DNA insert, at one site on the extent and patterns of changes in DNA methylation in the investigated *trans*-located sequences.

### *Stability of Transgene and Extent of Transgene Methylation*

Are strongly hypermethylated transgenes more stably integrated than hypomethylated ones? One approach to answer this question could be to genomically fix differently pre-methylated transgenes and follow their stability in individual cell clones.

### *Methylation of FV3 DNA*

This iridovirus is of obvious interest for studies on the interaction of specific proteins, particularly of transcription factors, with the fully 5'-CG-3'methylated viral genome in fish or mammalian cells. A major systematic approach on the biology and biochemistry of this viral infection will be required to understand the fundamental properties of this viral genome. Interesting new proteins might be discovered that interact with fully methylated viral DNA sequences both in fish and perhaps also in mammalian cells.

### *Methylation of Amplified 5'-$(CGG)_n$-3' Repeats in the Human Genome*

By what mechanism are amplified repeat sequences methylated? Could they be recognized as foreign DNA? A plasmid construct carrying increasing lengths of 5'-$(CGG)_n$-3' repetitions could be genomically fixed in the mammalian genome. In isolated clones of these cells, the extent of DNA methylation could be determined.

### *Infection of Epstein-Barr virus (EBV)-transformed Human Cells with Adenovirus: de Novo Methylation of Free Adenovirus DNA?*

DNA sequences in the persisting EBV genome can be methylated;

free adenovirus DNA in infected cells, however, remains unmethylated. The question arises as to whether free intranuclear adenovirus DNA in EBV-transformed cells can become de novo methylated in a nuclear environment in which DNA methyltransferases appear to be located also outside the nuclear chromatin, namely in association with the EBV genome.

## *Enzymes Involved in the de novo Methylation of Integrated Foreign DNA*

It is still uncertain which DNA methyltransferases or which combinations of these enzymes are involved in the de novo methylation of integrated foreign DNA. Enzyme concentration by itself might not be the rate-limiting step. Rather, chromatin structure and the topical availability of DNA methyltransferases could be the important factors that need to be investigated.

## *The Role of Specific RNAs in Triggering DNA Methylation*

There is a lack of studies on this problem in mammalian systems.

## *Complex Biological Problems Connected to DNA Methylation*

A great deal of very interesting research on DNA methylation derives from the work on epigenetic phenomena, on genetic imprinting, and more generally, from the fields of embryonal development, medical genetics, and tumor biology. From the currently available evidence, DNA methylation or changes in the original genomic patterns of DNA methylation are most likely implicated in any one of these phenomena. Current research, and examples of some of these investigations, are represented in these volumes, focusing on many of the highly complex details related to these problems. At present, we are undoubtedly still at the very beginning, and later editors of volumes in the series *Current Topics in Microbiology and Immunology* might help present progress in one or more of these exciting areas of molecular genetics.

# 2

# Replication and Translation of Epigenetic Information

Abstract Most cells in multicellular organisms contain identical genetic information but differ in their epigenetic information. The latter is encoded at the molecular level by post-replicative methylation of certain DNA bases (in mammals 5-methyl cytosine at CpG sites) and multiple histone modifications in chromatin. In addition, higher-order chromatin structures are generated during differentiation, which might impact on genome expression and stability. The epigenetic information needs to be "translated" in order to define specific cell types with specific sets of active and inactive genes, collectively called the epigenome. Once established, the epigenome needs to be "replicated" at each cell division cycle, i.e., both genetic and epigenetic information have to be faithfully duplicated, which implies a tight coordination between the DNA replication machinery and epigenetic regulators. In this review, we focus on the molecules and mechanisms responsible for the replication and translation of DNA methylation in mammals as one of the central epigenetic marks.

## INTRODUCTION

The term "post-genomic era," which is often used to classify the present scientific period, does not only stress the fact that the scientific community has finally reached beyond the mere deciphering of genomes, it also indicates that there is another level of genomic information apart from the one-dimensional nucleotide sequence. This *epi* (above/outside) genetic information is responsible for defining a cell type-specific state of the genome with a distinct set of active and inactive genes, the so-called epigenome. While the genome in a multicellular organism is identical for all cell types (with minor exceptions), the epigenome is potentially dynamic and cell type specific.

Epigenetic mechanisms have been reported to act by very different means, and an exhaustive description of the phenomenon is far from being completed. Some of these mechanisms act at the chromatin level as the methy-lation of DNA or the modification of histones by various functional groups including methyl, acetyl, phosphate, ADP-ribosyl groups or even such small proteins as ubiquitin or SUMO (reviewed, e.g., in Felsenfeld and Groudine 2003).

Other epigenetic modifications of chromatin include histone variants as well as chromatin-associated proteins like Polycomb group proteins. A different kind of epigenetic mechanism has been proposed to act at a global, topological scale, through the specific position of genes within the nucleus relative to functional nuclear subcompartments such as nucleoli, heterochro-matin, splicing compartments, etc. (reviewed, e.g., in: Cremer and Cremer 2001; Fisher and Merkenschlager 2002; Spector 2003). An emerging view is that the different epigenetic mechanisms can feedback onto each other, either strengthening a specific epigenetic state or weakening it, thereby enabling transition between transcriptionally permissive and repressive states of genes. In the present review we will address the propagation and translation of epigenetic information with the focus on DNA methylation in mammals.

## DNA Methylation

The modification of nucleotides in the DNA by covalently bound methyl groups was already described in the late 1940s and early 1950s (Hotchkiss 1948; Wyatt 1951). In the 1960s it was proposed that DNA methylation might be involved in a protection mechanism (1) against the integration of foreign DNA or (2) in rendering host DNA resistant to DNAses directed against foreign DNA (Srinivasan and Borek 1964). The latter idea went hand in hand with the discovery of bacterial restriction enzymes, which were thought to protect methylated bacterial host DNA from "invading" bacterial and viral DNA by specific digestion of the unmodified "parasitic" DNA (reviewed in Arber and Linn 1969).

It was not before 1975, though, that methylation of DNA in mammals was suggested to be connected with transcriptional regulation (Holliday and Pugh 1975; Riggs 1975).

DNA methylation is found in many different organisms including prokary-otes, fungi, plants, and animals, where it can serve different functions. Methyl groups in the DNA are found at the $C^5$ position of cytosines giving rise to 5-methyl cytosine (5mC) or at $N^6$ position of adenines resulting in $N^6$-methyladenine (6mA).

As already noted, methylation of DNA in bacteria is involved in a protection mechanism in which restriction endonucleases specifically digest foreign DNA by discriminating unmodified invader DNA sequences from methylated host DNA. In eukaryotic cells, the majority of methylated bases are cytosines, with only some, mostly unicellular organisms, showing low levels of methylated adenines (Gorovsky et al. 1973; Cum-mings et al. 1974; Hattman et al. 1978).

Methylation levels of eukaryotic DNA vary widely, from undetectable as in budding/fission yeast, nematodes or in adult *Drosophila melanogaster* flies over intermediate levels in mammals (2--8 mol%) up to high levels, reaching approximately 50 mol% in higher plants (see Doerfler 1983). In humans, approximately 1% of all DNA bases are estimated to be 5mC (Kriaucionis and Bird 2003). The sequence context in which methylated bases are found in eukaryotes is also variable. In mammals, for example, methylation is mainly found in CpG dinucleotides, with this "mini"-palindrome methylated on both strands. In fact 60%–90% of CpGs are methylated in mammalian genomes with the exception of so-called CpG islands, which are stretches of roughly 1 kb that frequently coincide with promoter regions.

These sequences, which are thought to be involved in transcriptional regulation, comprise roughly 1% of the mammalian genome. Exceptions to the rule that CpG islands are generally unmethylated are silenced genes on the inactive X-chromosome and at imprinted loci, where, depending on the parental origin, one allele is silenced. In contrast to mammals, methylation in fungi (reviewed in Selker 1997) and in plants (reviewed in Tariq and Paszkowski 2004) is not limited to CpG sites, with also CpNpG sequences being frequently methylated.

From an evolutionary point of view, DNA methylation is thought to represent an ancient mechanism, as the catalytic domain of DNA methyltransferases (Dnmts), the enzymes responsible for adding methyl groups to DNA, appears to be conserved from prokaryotes to humans (Kumar et al. 1994).

However, in the course of genome evolution there must have been adaptations concerninghow methyl marks were eventuallyutilized, since in different taxa DNA methylation appears to be involved in different functions. While in prokaryotes and fungi methylation appears mainly to serve protection needs of the host genome, in higher eukaryotes transcriptional silencing seems to be the main, though not the only, purpose.

A major change concerning the genomic organization as well as the extent of DNA methylation is thought to have occurred at the origin of vertebrate evolution, where DNA methyla-tion seems to have changed from a fractional organization, to a global one (Tweedie et al. 1997). In non-vertebrates, methylated DNA does not necessarily correlate with transposable elements or other functional chromosomal regions and appears not to be involved in transcriptional regulation, as no correlation could be found between transcription and methylation, neither for housekeeping genes, nor for tissue-specific genes (Tweedie et al. 1997).

In contrast, in mammals DNA methylation is implicated in many different aspects of transcriptional control including developmentally regulated genes, imprinted genes, and genes affected by X-inactivation. Nevertheless, it is also crucial for preventing spreading of potentially "parasitic" DNA elements like transposable sequences, thereby ensuring genome stability. Defects in DNA methylation have been shown to be involved in several pathological situations including cancer and other diseases such as Rett syndrome (RTT) or immunodeficiency, centromere instability, facial anomalies (ICF) syndrome. In the following sections, we will review two important aspects of DNA methylation, with an emphasis on the situation in mammals. In the first part we will reason how methylation marks are maintained in proliferating cells, i.e., how they are *replicated*, while in the second part we will concentrate on the question of how methylated CpGs are functionally interpreted in terms of transcriptional regulation, i.e., how the methyl cytosine information is *translated.*

## REPLICATION OF DNA METHYLATION

DNA methylation represents a post-synthetic modification, i.e., nucleotides are modified *after* they have been incorporated into the DNA. With respect to their substrate preference, two different kinds of Dnmts are distinguished:

(1) de novo Dnmts, which add methyl groups to completely unmethylated DNA and

(2) maintenance Dnmts that show a higher affinity for hemimethylated DNA, i.e., DNA where only one strand of the CpG palindrome is modified.

Hemimethylated DNA results from the replication of methylated regions. In both cases, the methyl-group donor is *S*-adenosyl-l-methionine (SAM). The three main, catalytically active Dnmts in mammals are Dnmt1, which is thought to serve as maintenance methyltransferase and Dnmt3a and 3b as de novo methylating enzymes.

Dnmt2 is expressed ubiquitously at low levels, but although it is among the most highly conserved Dnmts among different species all the way down to fission yeast, in most organisms it could not yet be shown to possess catalytic activity (Okano et al. 1998; Yoder and Bestor 1998; discussed in Robertson 2002). In *D. melanogaster,* however, Dnmt2 is responsible for the low level DNA methylation found during embryonic stages (Kunert et al. 2003). Due to its evolutionary conservation, Dnmt2 might well represent the ancestral Dnmt protein.

The de novo methylating enzymes Dnmt3a and 3b are supposed to be responsible for methylation of the embryonic genome after implantation, i.e., after the parental genomes have been demethylated (Okano et al. 1999). Dnmt3a and Dnmt3b have been shown to be catalytically active in vitro as well as in vivo, and transcripts were found in embryonic stem (ES) cells, in the early embryo as well as in adult tissue and in tumor cells (see citations in Robertson et al. 1999). Two isoforms of Dnmt3a were described, one reported to bind euchromatin and the other heterochromatin (Okano et al. 1998; Chen et al. 2002).

*Dnmt3a* knockout ES cell lines appeared to be normal concerning their de novo methylation potential, and null mice developed inconspicuously until birth, but shortly after showed decreased growth and died by 4 weeks of age (Okano et al. 1999). Dnmt3b shows at its N-terminus only little sequence homology to Dnmt3a, and unlike Dnmt3a, its expression is low in most tissues, but high in testis, so that an implication in methylation during spermatogenesis has been proposed (Okano et al. 1998; Robertson et al. 1999; Xie et al. 1999). Its localization in centromeric regions in ES cells (Bachman et al. 2001) and the observation that mutant *Dnmt3b*$^{-/-}$ cells exhibit a decreased methylation of minor satellite repeats (Okano et al. 1999) suggested a role in centromeric satellite methylation. Dnmt3b appears to be more important during embryonic development than Dnmt3a, since no viable null mice were obtained (Okano et al. 1999). Mutations in *Dnmt3b* in humans cause so-called ICF syndrome, where pericentric repeats are hypomethylated (Hansen et al. 1999; Okano et al. 1999; Xu et al. 1999). Several Dnmt3b splicing isoforms have been found. The eight variants described in mouse and the five in humans are expressed in a tissue-specific manner, yet not all of them appear to be catalytically active.

Within the Dnmt3 family but more distantly related is the Dnmt3L protein that lacks the conserved motifs of C5-methyltransferases and was found to be highly expressed in mouse embryos and testis (Aapola et al. 2001). *Dnmt3L* null mice show methylation defects at maternal imprints (Bourc'his et al. 2001) but otherwise a normal genome-wide methylation pattern, which suggests that Dnmt3L is involved in the establishment of maternal imprints, probably by recruiting Dnmt3a or 3b to target loci, either directly or indirectly.

The Dnmt1 enzyme was the first to be cloned (Bestor et al. 1988) and was shown to be essential for development, since null mice die at mid-gestation (Li et al. 1992). Interestingly *Dnmt1*$^{-/-}$ ES cells are viable and show normal morphology and a 5mC level that is still 30% of that in wild-type cells, suggesting some compensatory methylation activity (Li et al. 1992), likely due to Dnmt3a/3b enzymes. Across various mammalian species, the N-terminus of Dnmt1 appears to be rather variable, while the catalytic C-terminus is more conserved (Margot et al. 2000). The intracellular distribution of Dnmt1 is rather dynamic throughout the cell cycle. The enzyme is diffusely distributed throughout the nucleoplasm during most of G1, associates with subnuclear sites of DNA replication during S-phase (Leonhardt et al. 1992), and binds to chromatin, with preference to pericentric heterochromatin, duringG2 andM-phases (Easwaran et al. 2004). This complex cell cycle distribution of Dnmt1 has also been exploited to construct cell-cycle marker systems (Easwaran et al. 2005).

Since Dnmt1 messenger (m)RNA has also been found in low proliferative tissue (Robertson et al. 1999), where only few cells are suspected to be actually replicating DNA, it has been proposed that Dnmt1 might exert an additional function beyond methylating hemimethylated DNA during S-phase. In fact, isoforms of Dnmt1 have been found that could account for additional functions. The originally cloned Dnmt1 (Bestor et al. 1988) was found later to be missing a 118 amino acid sequence at its N-terminus (Tucker et al. 1996; Yoder et al. 1996). This longer Dnmt1 protein (Dnmt1L; 1,620 amino acids) is expressed in most proliferating somatic cells, while the original shorter Dnmt1 protein (Dnmt1S; 1,502 amino acids) accumulates specifically during oocyte growth (Mertineit et. al. 1998).

While at the protein level, two forms are known, at the mRNA level, three isoforms with differingfirst exons/promoters have been described. In addition to the predominant somatic isoform, two sex-specific isoforms were isolated. One isoform is the only one expressed in oocytes and corresponds at the protein level to the shorter form (Mertineit et al. 1998). It localizes in the cytoplasm of mature oocytes, except for the 8-cell stage, where it is transiently relocated into the nucleus (Carlson et al. 1992; Cardoso and Leonhardt 1999). Since knockout female but not male mice were infertile, with embryos from deficient females showing defective methylation pattern at imprinted loci, the current idea is that oocyte Dnmt1 and especially its nuclear localization at the 8-cell stage isimportant formaintainingimprints (Howell et al. 2001).

During mouse preimplantation development, while the genome is globally demethylated, this Dnmt1 form appears to be responsible for keeping the retrotransposable element IAP (intracisternal A-type particle) methylated and thus silent (Gaudet et al. 2004). Silencing of such mobile elements is thought to be crucial to prevent transcriptional activation and potential mutagenesis by transposition. The second sex-specific isoform was originally detected in pachytene spermatocytes (Mertineit et al. 1998). The same isoform, however, was found also in differentiated myotubes, instead of the ubiquitously expressed Dnmt1, which is downregulated upon differentiation (Aguirre-Arteta et al. 2000). Since myotube nuclei show no DNA replication, this isoform might serve a function that is independent of DNA synthesis. Both oocyte and spermatocyte/skeletal muscle mRNA isoforms give rise to the shorter Dnmt1 protein form.

The marked preference of Dnmt1 for hemimethylated DNA together with its specific association withreplication machinery during S-phase via binding to proliferating-cell nuclear antigen (PCNA) (Leonhardt et al. 1992; Chuang et al. 1997; Easwaran et al. 2004) make it a strong candidate for mediating the propagation of the DNA methylation pattern at each cell division cycle. Since Dnmt1 is a catalytically slow enzyme (Pradhan et al. 1997), its prolonged association in G2 and M-phases with chromatin could allow sufficient time for full methylation of all hemimethylated sites, in particular at heavily methylated heterochromatic sequences (Easwaran et al. 2004). In addition, Dnmt1 has been reported to interact with histone deacetylases (HDACs) (Fuks et al. 2000; Robertson et al. 2000; Rountree et al. 2000) and might serve as a loading platform for these chromatin modifiers.

Concomitantly, methyl-CpG-binding domain (MBD) proteins, recognizing the newly generated modified CpGs, have been also shown to recruit HDACs (Jones et al. 1998; Nan et al. 1998; Ng et al. 1999) and can thereby further contribute to the replication of the histone modifications upon DNA replication. In this regard, there is increasing evidence of crosstalk between histone modifications and DNA methylation.In paralleltothesemechanismsfor replicationofepigenetic information, the random distribution of "old" histones between the two replicated DNA strands implies that modifications such as histone methylation are passed onto the nucleosomes assembled at the newly replicated strands. Factors such as HP1, which recognizes specific methylation forms of histone H3 (Lachner et al. 2001), can then bind the replicated chromatin, recruit histone methyltransferases (HMTs)

(Lehnertz et al. 2003) and "spread" the histone methylation marks onto the adjacent, previously deacetylated histones.

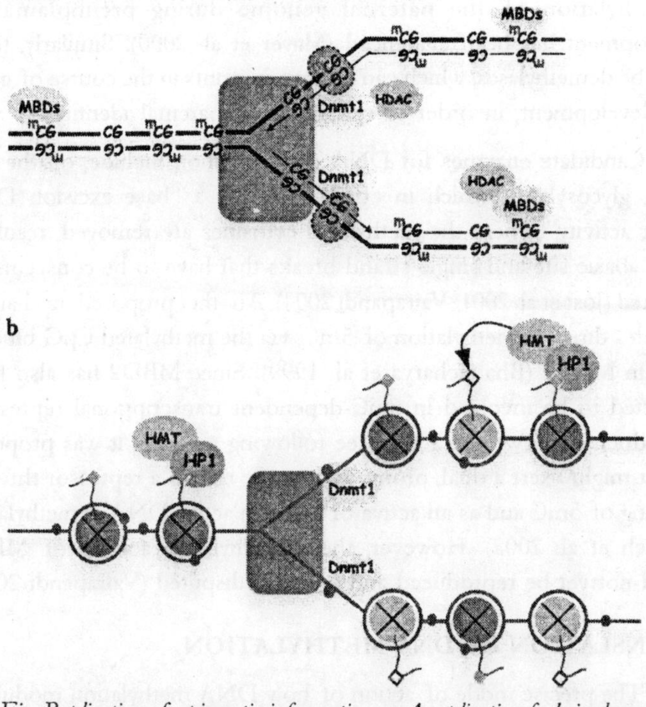


*Fig. Replication of epigenetic information. a A replication fork is shown where Dnmt1 associated with the replication machinery* (green box*) is copying the methylation mark* (m*) at hemimethylated CpG sites, which are then recognized and bound by methyl-CpG-binding domain (MBD) proteins. Both MBD proteins and Dnmt1 recruit histone deacetylases (HDACs), thereby maintaining the deacetylated chromatin state. b The same replication fork is shown from a nucleosomal view. Nucleosomes are shown as* blue circles, *with methylated histone H3 tails as* filled yellow squares *and 5mC as*red dots. *Histones bearing repressive methylated lysine residues are distributed randomly onto replicated daughter strands. Binding of HP1 to methylated histones can recruit histone methyltransferase (HMT) that modify lysine residues of the newly incorporated histones (*light blue circles*)*

Although many enzymes have been described that can actually add methyl groups to the DNA, much less is known about DNA demethylases.

The existence of such enzymes, however, is almost certain, since active demethylation of the paternal genome during preimplantation development has been evidenced (Mayer et al. 2000). Similarly, there must be demethylases, which can remove imprints in the course of germ cell development, in order to set the novel parental identity.

Candidate enzymes for DNA demethylation include, on the one hand, glycosylases, which in effect resemble a "base excision DNA repair activity" where the methylated cytosines are removed, resulting in an abasic site and single strand breaks that have to be consecutively repaired (Jost et al. 2001; Vairapandi 2004). Another proposed mechanism includes direct demethylation of 5mC, via the methylated CpG binding protein MBD2 (Bhattacharya et al. 1999). Since MBD2 has also been reported to be involved in 5mC-dependent transcriptional repression (Hendrich and Tweedie 2003) (see following section), it was proposed that it might exert a dual, promoter-specific role as a repressor through binding of 5mC and as an activator through active DNA demethylation (Detich et al. 2002). However, the demethylating activity of MBD2 could not yet be reproduced and is hence disputed (Vairapandi 2004).

## TRANSLATION OF DNA METHYLATION

The precise mode of action of how DNA methylation modulates transcription is far from being understood. In fact, different mechanisms could account for controlling gene expression at different loci. Though DNA methylation in general is associated with transcriptional silencing, in some cases methy-lation has been shown to induce expression. This has been demonstrated for the imprinted *Igf2* locus, where methylation of a differentially methylated region (DMR) on the maternal chromosome prevents binding of CTCF (CCCTC-binding factor), which results in a positive enhancer function (Bell and Felsenfeld 2000; Hark et al. 2000; Kanduri et al. 2000; Szabo et al. 2000). Transcriptional silencing mediated by methylation of CpGs near promoter regions is thought to occur by at least two different mechanisms. One possibility is that methylation of specific target sites simply abolishes binding of transcription factors or transcriptional activators by sterical hindrance. Another increasingly important mechanism involves the specific recognition and binding of factors to methylated DNA, triggering different kinds of downstream responses, entailing (or not) further chromatin modifications. In mammals, there are several known methyl-CpG-binding proteins.

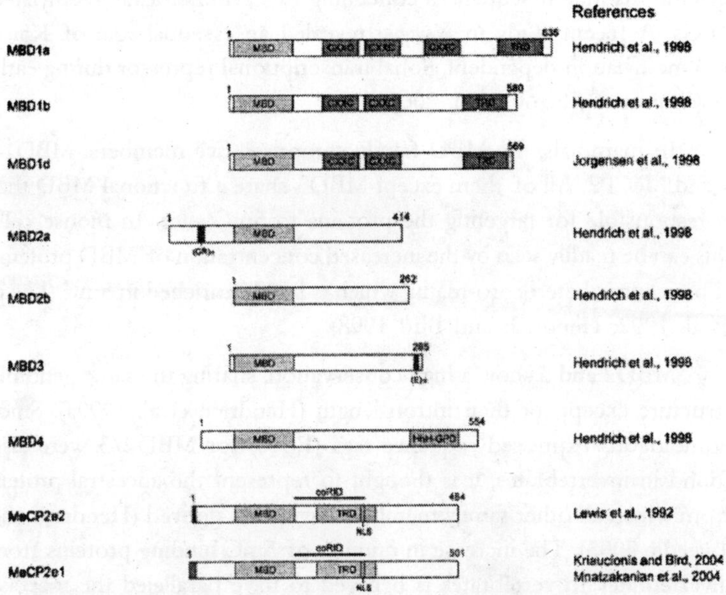


*Fig. Organization of the mouse MBD protein family.* **Numbers** *represent amino acid positions.* **coRID,** *corepressor interacting domain;* **CXXC,** *Cys-rich domain;* **$(E)_{12}$,** *Glu repeat;* **$(GR)_{11}$,** *Gly-Arg repeat;* **MBD,** *methyl-CpG-binding domain;* **HhH-GPD,** *DNA N-Glycosylase domain;* **TRD,** *transcriptional repressor domain*

The MBD protein familymembers share a conserved*m*ethyl-CpG-*b*inding*d*omain (MBD)(HendrichandBird1998).While MeCP2, MBD1,andMBD2 have been shown to act as transcriptional repressors, MBD4 appears to be involved in reducing the mutational risk from potential C'!T transitions, which result from deamination of 5mC. A fifth member of the MBD family, MBD3 does not bind to methylated DNA (Hendrich and Tweedie 2003), but is a constituent of the NuRD (nucleosome remodeling and histone deacetylation) corepressor complex. A further, recently detected 5mC-binding protein is Kaiso, which shows no sequence conservation with MBD proteins but also functions as a transcriptional repressor (Prokhortchouk et al. 2001). In contrast to

MBDs, Kaiso appears to bind via a zinc-finger motif in a sequence-specific manner at sequences containing two symmetrically methylated CpGs. A recent study in *Xenopus* revealed an essential role of Kaiso as a methylation-dependent global transcriptional repressor during early development (Ruzov et al. 2004).

In mammals, the MBD family comprises five members: MBD1–4 and MeCP2. All of them except MBD3 share a functional MBD that is responsible for targeting the proteins to 5mC sites. In mouse cells this can be readily seen by the increased concentration of MBD proteins at pericentric heterochro-matin, which is highly enriched in 5mC (Lewis et al. 1992; Hendrich and Bird 1998).

MBD2 and 3 show a high conservation, sharing the same genomic structure except for their intron length (Hendrich et al. 1999a). Since homologous expressed sequence tags (ESTs) for MBD2/3 were also found in invertebrates, it is thought to represent the ancestral protein from which all other family members have been derived (Hendrich and Tweedie 2003). The increase in number of 5mC binding proteins from invertebrates to vertebrates is believed to have paralleled the increase in DNA methylation, as this would have enabled a fine-tuning of methylation-dependent silencing on the one hand, as well as lowered the mutational risks emerging from spontaneous deamination on the other (Hendrich and Tweedie 2003).

In mammals, MBD3 does not bind to methylated CpGs due to two amino acid substitutions within the MBD (Saito and Ishikawa 2002). Other vertebrates, however, such as frogs, have two MBD3 forms, one of which retains a 5mC-binding ability (Wade et al. 1999). Sequence homology predicts a similar situation for the pufferfish and the zebrafish (Hendrich and Tw e e die 2003).

MBD3 in mammals is a constituent of the NuRD corepressor complex. NuRD is found in many organisms including plants and plays an important role in transcriptional silencing via histone deacetylation. Though MBD3 has been shown to be essential for embryonic development (Hendrich et al. 2001), its function within the NuRD multiprotein complex has still to be clarified. MBD2 interacts with the NuRD complex making up the MeCP1 complex (methyl-CpG-binding protein), which was actually the first methyl-CpG-binding activity isolated in mammals (Meehan et al. 1989). In spite of the many potential binding sites of MBD2, it does not appear to act as a global transcriptional repressor.

In fact, only one target gene of MBD2 has been described until now, and that is *Il4* during mouse T cell differentiation (Hutchins et al. 2002). Here loss of MBD2 has been shown to correlate with a leaky instead of a complete repression. Consequently, it has been hypothesized that MBD2 might rather act in "fine-tuning" transcriptional control by reducing transcriptional noise at genes, which are already shut off (Hendrich and Tw e e die 2003). Alternatively, the lack of a global de-repression of methylated genes upon MBD2 loss could be explained by redundancy among MBD family members. Studies abrogating several MBD proteins at the same time will help to answer this question. An interesting phenotype of MBD2$^{-/-}$ mice is that affected female animals neglect their offspring due to an unknown neurological effect (Hendrichetal. 2001). MBD2bis anisoform that is generatedbyusingan alternative translation start codon generating a protein that lacks 140 N-terminal amino acids (Hendrich and Bird 1998). Surprisingly, it has been reported to possess a demethylase activity (see previous section and Bhattacharya et al. 1999). In gene reporter assays, it was even shown to act as a transcriptional activator (Detich et al. 2002). Thus, it has been proposed that MBD2 could act as both a transcriptional repressor and stimulator. It should be added, though, that other groups have not been able to reproduce the demethylase activity of MBD2b, so the existence of this activity is still controversial (discussed in Wa de 2001).

MBD1 is exceptional among the transcriptionally repressive MBDs, since it can suppress transcription from both methylated and unmethylated promoters in transient transfection assays (Fujita et al. 1999). Four splicing isoforms have been described in humans (Fujita et al. 1999) and three in mouse (Jor-gensen et al. 2004), with the major difference being the presence of three versus two CXXC cysteine-rich regions (see Fig. 3). The presence of the most C-terminal CXXC motifs in mouse was shown to be responsible for its binding to unmethylated sites (Jorgensen et al. 2004) and for its capacity to silence un-methylated reporter constructs (Fujita et al. 1999). The repression potential of

MBD1 seems to rely on the recruitment of HDACs, although, most probably, different ones from those engaged in MBD2 (and MeCP2) silencing (Ng et al. 2000). Similar to MBD2, MBD1$^{-/-}$ mice exhibit neurological deficiencies, as they show reduced neuronal differentiation and have defects in spatial learning as well as in hippocampus long-term potentiation (Zhao et al. 2003).

MBD4 is the only member within the MBD family that is not involved in transcriptional regulation. Instead, it appears to be implicated in reducing the mutational risk that is imminent in genomes with high methylation levels, by transitions of 5mC'!T via deamination. This transition poses a bigger problem for the DNA repair machinery than C'!U transitions, which result from the deamination of unmethylated cytosines, since the former results in G–T mismatches, in which the mismatched base (G or T) cannot readily be identified. In contrast, uracil in G–U mismatches can easily be pinpointed as the "wrong" base, since it is not a constituent of DNA. Accordingly, MBD4 possesses a C-terminal glycosylase moiety that can specifically remove Ts from G–T mismatches (Hendrich et al. 1999b; see Fig. 3). In fact, its preferred binding substrate is 5mCpG/TpG, i.e., the deamination product of the 5mCpG/5mCpG dinucleotide. Indeed mutation frequency analysis in MBD4$^{-/-}$ mice revealed an approximately threefold increase in C'!T transitions at CpGs compared to wild-type cells (Millar et al. 2002; Wong et al. 2002), which supports the idea of MBD4 being a mutation attenuator.

Since MeCP2 was the first methyl-CpG-binding protein to be cloned and the second methylated DNA binding activity to be isolated after MeCP1, it is often referred to as the founding member of the MBD family. A single methylated CpG dinucleotide has been shown to be sufficient for binding (Lewis et al. 1992). In transient transfection assays with methylated gene reporter in *Xenopus* and in mice it was demonstrated that MeCP2 functions as a transcriptional repressor, at least in part via interaction with the Sin3 corepressor complex, which contains histone deacetylases 1 and 2 (Jones et al. 1998; Nan et al. 1998). An approximately 100-amino-acid-containing tran-scriptional repression domain (TRD) in the middle of the protein has been shown to be critical for transcriptional silencing (Nan et al. 1997). Apart from the recruitment of HDACs, MeCP2 has been shown to associate with a histone methyltransferase activity specifically modifyinghistone H3 at lysine 9, which is known to represent a transcriptionally repressive chromatin label (Fuks et al. 2003). In addition, MeCP2 has recently been found to interact with components of the SWI/SNF-related chromatin-remodeling complex, suggesting a novel potential MeCP2-dependent silencing mechanism (Harikrishnan et al. 2005). Moreover, MeCP2 can induce compaction of oligonucleosomes in vitro, which could additionally suppress transcription in vivo through a dense chromatin conformation that is incompatible with the binding of factors relevant for transcriptional activation (Georgel et al. 2003). In summary, MeCP2 could translate the DNA methylation mark directly by preventing the access of transcriptional activators to promoters/enhancers or indirectly by either recruiting modifiers of histones such as histone deacetylases and methyltransferases or by compacting chromatin.

With the idea in mind that MeCP2 might act as a global transcription repressor, it was very surprising that an expression profiling analysis comparing MeCP2 null mice with normal animals revealed only subtle changes in the mRNA profiles of brain tissues (Tudor et al. 2002). This apparent lack of global de-repression in the absence of MeCP2 resembles a similar situation as described for MBD2$^{-/-}$ mice (as discussed earlier in this section). Possible reasons for this observation could be either that other MBD proteins can compensate for the loss of MeCP2, or that the changes in transcription levels induced by MeCP2 deficiency are so small that they are undetectable with current microarray technology. This supports the rationale that MBDs might act as reducers of transcriptional noise rather than to shut down active genes (Hendrich and Tw e e die 2003).

On the other hand, it could well be that MeCP2 represses genes in a tissue- and/or time-specific fashion. Matarazzo and Ron-nett, for example, using a proteomic approach, found substantial differences in protein levels between MeCP2-deficient and wild-type mice (Matarazzo and Ronnett 2004). Importantly, they showed that the degree of differences varied depending on the analyzed tissue (olfactory epithelium vs olfactory bulb) and the age of the animals (2 vs 4 weeks after birth). Apart from a potential global effect, MeCP2 has recently been linked to the regulation of two specific target genes. The genes of Hairy2a in *Xenopus* (Stancheva et al. 2003) and brain-derived neurotropic factor (BDNF) in rat (Chen et al. 2003) and mice (Martinowichet al. 2003)—bothare proteins involvedin neuronal development and differentiation—have methylated promoters with bound MeCP2, which is released upon transcriptional activation. Recently MeCP2 was shown to be involved in the transcriptional silencing of the imprinted gene *Dlx5* via the formation of a chromatin loop structure (Horike et al. 2005).

MeCP2 is expressed ubiquitously in many tissues of humans, rats, and mice, although at variable levels. Several lines of evidence argue that MeCP2 expression increases during neuronal maturation and differentiation (Shah-bazian et al. 2002b; Jung et al. 2003; Balmer et al. 2003; Cohen et al. 2003; Mullaney et al. 2004). In a recent study, it was shown that MeCP2 and MBD2 protein levels increase also during mouse myogenesis along with an increase in DNA methylation at pericentric heterochromatin (Brero et al. 2005). Moreover, it was demonstrated that MeCP2 and MBD2 are responsible for a major reorganization of pericentric heterochromatin during terminal differentiation that leads to the formation of large heterochromatic clusters (Brero et al. 2005). This finding provides the link between a protein(s) (MeCP2/MBD2) and chromatin organization and assigns it a direct role in changes of the 3D chromatin topology during differentiation. The latter represents yet another level of epigenetic information beyond the molecular composition of chromatin.

In agreement with its substrate specificity, MeCP2 localizes mainly at heavily methylated DNA regions. In mouse nuclei, for example, MeCP2 intensely decorates pericentric heterochromatin (Lewis et al. 1992). In human cells, however, the intranuclear distribution of MeCP2 was found to deviate from the pattern in mouse, in that it did not strictly colocalize with methylated DNA, pericentric satellite sequences, or heterochromatic regions [visualized by intense 4′-6'-diamidino-2-phenylindole (DAPI) staining; Koch and Stratling 2004]. Intriguingly, the authors found an additional binding affinity of MeCP2 for TpG dinucleotides and proposed a sequence-specific binding defined by adjacent sequences. By using an immunoprecipitation approach, they revealed an association of MeCP2 with retrotransposable elements, especially with Alu sequences, and with putative matrix attachment regions (MARs).

In this respect, it should be added that the MeCP2 homolog in chicken (named ARBP) was originally isolated as a MAR binding activity (von Kries et al. 1991), even before rat MeCP2 was actually described for the first time (Lewis et al. 1992), yet its homology to the rat protein was noticed only later (Weitzel et al. 1997). Interestingly, ARBP/MeCP2 binding in chicken appears not to be dependent on CpG methylation (Weitzel et al. 1997). Since the results in human cells were obtained using a breast cancer cell line (MCF7), it will be interesting to investigate further human cell types, including primary cells, to further clarify MeCP2 binding specificity in human cells.

Two studies have lately reported a second MeCP2 splicing isoform, which yields a protein with a slightly different N-terminal end, due to the utilization of an alternative translation start codon (Kriaucionis and Bird 2004; Mnatzakanian et al. 2004; Fig. 3). Surprisingly this new MeCP2 mRNA appears to be much more abundant in different mouse and human tissues than the originally described isoform. Fluorescently tagged fusions of both proteins, though, show the same subnuclear distribution in cultured mouse cells (Kriaucionis and Bird 2004). An antibody raised against the "old" isoform was shown to recognize also the novel variant (Kriaucionis and Bird 2004). Consequently, in previous immunocytochemical studies most probably both isoforms have been detected. The differences between both isoforms are only subtle, with the new protein having a 12 (human) and 17 (mouse) amino acid longer N-terminus followed by a divergent stretch of 9 amino acids. Since neither the MBD nor the TRD are affected by the changes, both proteins are anticipated to be functionally equivalent.

As already noted, MeCP2 expression appears to be correlated with differentiation and development. Its implication in neuronal differentiation is further supported by its involvement in a human neurodevelopmental disorder called Rett syndrome (RTT). The syndrome was originally described in 1966 by the Austrian pediatrician Andreas Rett, but its genetic basis was revealed only recently (Amir et al. 1999). At least 80% of RTT cases are caused byspontaneous mutations in the MeCP2 gene (see Kriaucionis and Bird 2003), which is localized on

Xq28 (Amir et al. 1999). RTT is the second most frequent form of female mental retardation after Down syndrome, and its incidence is approximately twofold higher than phenylketonuria (Jellinger 2003). RTT is diagnosed in 1:10,000–1:22,000 female births, with affected girls being heterozygous for the *MeCP2* mutation (Kriaucionis and Bird 2003); consequently, the phenotype is caused by the cells that do not express functional protein due to random inactivation of the X chromosome containing the wild-type copy of *MeCP2*. Most mutations found in RTT patients are located within the functional domains, i.e., within the MBD and the TRD of MECP2, but several mutations have also been found in the C-terminal region, where no concrete function has yet been assigned.

Recently, however, it was shown that the C-terminal domain of MeCP2 is crucial at compacting oligonucleosomes into dense higher order conformations in vitro (Georgel et al. 2003). Interestingly, this activity was found to be independent of CpG methylation of the oligonucleosomal arrays, which parallels the findings in human and chicken where MeCP2 binding was also found at non-methylated sites (see above) (Weitzel et al. 1997; Koch and Stratling 2004). Moreover, the C-terminal domain of MeCP2 was found to specifically bind to the group II WW domain found in the splicing factors formin-binding protein (FBP) and HYPC (Buschdorf and Stratling 2004). Although the functional role of this association has yet to be unraveled, various mutations within this C-terminal region were shown to correlate with a RTT phenotype.

In mouse models for RTT, animals carrying mutations in the C-terminus generally exhibit a less-severe phenotype than those with a null mutation (Shahbazian et al. 2002a). Mice where MeCP2 was conditionally knocked out only in brain tissue yielded the same phenotype as that where the whole animal was affected, suggesting that the observable phenotype is largely due to a failure of proper brain development (Chen et al. 2001; Guy et al. 2001). Mutations in MeCP2, moreover, have been shown to correlate with phenotypes containing clinical features of X-linked mental retardation (Couvert et al. 2001), Angelman syndrome (Watson et al. 2001), and autism (Carney et al. 2003; Zappella et al. 2003). In conclusion, RTT is a good example illustrating that not only are the establishment and replication of methylation marks pivotal for a normal development—as is shown by the severe pheno-types caused by loss of Dnmt functions—but the correct translation of DNA methylation marks is a critical prerequisite for normal ontogeny.

## Outlook

The establishment andstable maintenance of epigenetic marks on thegenome at each cell division as well as the translation of this epigenetic information into genome expression and stability is crucial for development and differentiation. This role of epigenetic regulatory mechanisms in the realization of the genome has been clearly established by the finding of mutations affecting epigenetic regulators in human diseases (RTT and ICF syndrome) and the severity of phenotypes in animal models carrying mutations in the different components of these pathways. In addition, global and local changes in methylation patterns of the genome are found in most tumors and have, therefore, triggered intense research into their usage as new tumor diagnostic tools and therapeutic targets.

Another recently emerging and exciting area of research where manipulating epigenetic information is of fundamental importance is stem cell therapy and animal cloning. In a reversed way to differentiation, resetting or repro-gramming of the epigenetic state of a differentiated donor cell appears to be one of the major difficulties in animal cloning by nuclear transfer (reviewed, e.g., in Shi et al. 2003). Besides having a fundamental impact for basic research, understanding the nature of epigenetic information and its plasticity in (adult/embryonic) stem cells is a key prerequisite for successful clinical applications of cell replacement therapies in regenerative medicine.

# 3

# DNA Methyltransferases: Facts, Clues, Mysteries

## INTRODUCTION

DNA methylation is a major epigenetic event. It is a post-replicative, reversible, and heritable chemical modification of DNA involved in regulating a diverse range of biological processes in vertebrates, plants, and fungi. The present chapter deals mainly with DNA methylation in mammals, and particularly humans and mice.

In mammals, DNA methylation occurs predominantly at cytosine residues located within CpG dinucleotides and is associated with gene silencing. The distribution of CpG dinucleotides in the mammalian genome is uneven and non-random.

Methylated DNA is most abundant in heterochromatin-containing bulk DNA such as parasitic sequences, retrotransposons, and various repeat elements. Most unmethylated CpG dinucleotides are found in "CpG islands," i.e., small stretches of CpG-rich DNA found in the 5' regulatory regions of almost half of the genes of the genome (Bird 2002).

DNA methylation has a crucial role in normal mammalian development and plays a major role in gene expression, X-chromosome inactivation in females, and genomic imprinting. It also contributes to the stability and integrity of the genome by inactivating bulk DNA. Altered methylation patterns, with genome-wide hypomethylation and region-specific hypermethylation, are frequently found in cancers (Jones and Baylin 2002).

How does DNA methylation lead to gene silencing? How are DNA methylation patterns established and maintained? These are among the most pressing and intriguing questions in the DNA methylation field. Mechanistic insights into these questions have come from the identification and characterization of several dedicated enzymes called DNA methyltransferases (DNMTs). These key regulators of DNA methylation are the focus of this chapter, which leads the reader on a trail that starts with the structure of these proteins and progresses through the mechanisms by which they repress transcription and what we know about their targeting to preferred DNA sequences. Emphasis is laid on emerging evidence of an intimate connection between DNMTs and chromatin structure.

## DNMTs: Mug Shots and Knockout

DNMTs catalyze methylation at position 5 of the cytosine ring, using *S*-adenosyl-methionine as the methyl group donor. On the basis of sequence homology, DNMTs are divided into three families: DNMT1, DNMT2, and DNMT3. This third family has three members: DNMT3A, DNMT3B, and DNMT3L. The structures and enzymatic activities of these proteins and the corresponding knockout phenotypes are reviewed in the following sections.

### DNMT STRUCTURE

A DNMT generally comprises two domains: a well-conserved catalytic domain in the carboxy-terminal part of the protein and a more variable regulatory domain in the amino-terminal region. Dnmt1 was the first enzyme to be isolated as a mammalian DNMT and the only one identified via a biochemical assay (Bestor et al. 1988; Yen et al. 1992). It has the largest amino-terminal domain of all known DNMTs. Responsible for import into the nucleus and for zinc binding, this domain also mediates protein-protein interactions.

Expression of the gene *Dnmtl* is high in proliferating cells and ubiquitous in somatic cells. During gametogenesis, expression of the gene from sex-specific promoters and 5' exons results in sex-specific Dnmt1 isoforms whose biological functions are still quite obscure (Mertineit et al. 1998; Doherty et al. 2002). In the mouse, a Dnmt1 isoform called Dnmt1o, for "oocyte-specific," is expressed in the oocyte and pre-implantation embryo. It seems to be required only during a single S-phase in the 8-cell mouse embryo to maintain methylation patterns at imprinted loci (Howell et al. 2001).

The observation that methylation persists in mouse embryonic stem cells lacking the *Dnmtl* gene led researchers to postulate that other DNMTs must exist. Screening of expressed sequence tag (EST) databases for sequences containing motifs of the conserved catalytic domain led to the identification of three candidates: Dnmt2, Dnmt3a, and Dnmt3b (Okano et al. 1998a; Yoder and Bestor 1998).

Dnmt2 contains only the DNMT motifs; its gene is expressed, albeit to low levels, in many human and mouse tissues (Yoder and Bestor 1998). The role of this protein remains enigmatic.

The genes *Dnmfia* and *Dnmfib* show very high expression during em-bryogenesis and gametogenesis but much lower expression in differentiated somatic tissues.

Two Dnmt3a and seven Dnmt3b isoforms have been described, featuring specific expression patterns during development and in adult tissues. Very little is known about the biological importance of individual isoforms (Okano et al. 1998a; Chen et al. 2002). To elucidate the specific function of each Dnmt3a and Dnmt3b isoform, it will be necessary to carry out genetic analyses based on isoform-specific gene disruption.

Structurally, Dnmt3a and Dnmt3b share, in addition to the catalytic site in the C-terminal region, two conserved domains in the amino-terminal region: the proline- and tryptophan-rich PWWP domain and the cysteine-rich PHD domain (for plant homeodomain).

The PWWP domain has been found in more than 60 eukaryotic proteins implicated in transcriptional regulation and chromatin organization (Stec et al. 2000).

The structure of the mouse Dnmt3b PWWP domain is known (Qiu et al. 2002). This domain probably allows targeting of Dnmt3a and Dnmt3b to pericentric heterochromatin, as it is sufficient for binding to metaphase chromosomes and promotes methylation of nucleosomal DNA (Chen et al. 2004; Ge et al. 2004). The PWWP domain of Dnmt3b binds nonspecifically to DNA (Qiu et al. 2002); that of Dnmt3a shows little DNA-binding ability (Chen et al. 2004).

The second conserved domain of the N-terminal region, the PHD domain, is conserved also in the third member of the DNMT3 family, Dnmt3L. The PHD domains of these proteins most closely resemble the imperfect PHD motif found in ATRX, a putative member of the SNF2 family of AT P- dependent chromatin remodeling proteins. A mutated ATRX gene has been found in several X-linked mental retardation disorders (Gibbons et al. 2000). The PHD domain mediates protein–protein interactions and functions as a transcrip-tional repressor domain (Burgers et al. 2002).

## *Dnmt* Knockout in Mice

DNA methylation changes in a highly orchestrated way in the course of mouse development. This involves both genome-wide and gene-specific demethyla-tion and de novo methylation (Li 2002). As mentioned above, DNA methyla-tion is essential to mammalian development. This is vividly illustrated by targeted disruptionofDNMTgenesinmice,whichcausesembryonic (*Dnmt1*and *Dnmt3b*) or post-natal (*Dnmt3a*) mortality (Li et al. 1992; Okano et al. 1999).

*Dnmt1*$^{-/-}$ mice die around embryonic day (E)8.5, at the onset of gastrula-tion. Analyses of dead embryos have revealed genome-wide demethylation, biallelic expression of several (but not all) imprinted genes, and aberrant expression of Xist, a long, non-coding RNA involved in X-chromosome inac-tivation in females (Li et al. 1992).

*Dnmt3a*$^{-/-}$ mice die 4 weeks after birth; they display severe intestinal defects and impaired spermatogenesis. As for *Dnmt3b*$^{-/-}$ mice, they show demethylation of minor satellite DNA, mild neural tube defects, and embryo mortality at E14.5–E18.5 (Okano et al. 1999). When both *Dnmt3a*and *Dnmt3b* are disrupted in mice, doubly homozygous [*Dnmt3a*$^{-/-}$, *Dnmt3*b$^{-/-}$] embryos have a phenotype similar to that of *Dnmt1*$^{-/-}$ embryos, showing developmental arrest at the presomite stage and a distorted neural tube around E8.5 (Okano et al. 1999).

Mice withadisrupted*Dnmt2*gene are viable and fertile, withminor defects (Okano et al. 1998b). This is in agreement with results obtained on *Dnmt2*$^{-/-}$ embryonic stem (ES) cells. These cells are viable and show no obvious alteration of their DNA methylation pattern (Okano et al. 1998b). As mentioned in the next section, this mild phenotype of *Dnmt2*$^{-/-}$ is probably linked to the very low enzymatic activity of the DNMT2 protein (Hermann et al. 2003). *Dnmt3L*$^{-/-}$ mice are viable, but males are sterile and the heterozygous progeny of homozygous females die in utero and show complete loss of maternal genomic imprinting (Hata et al. 2002). This phenotype is indistinguishable from that of conditional knockout mice having a disrupted *Dnmt3a* gene in germ cells only. This highlights the crucial role of Dnmt3L and Dnmt3a in maternal imprinting (Kaneda et al. 2004). A study also suggests that Dnmt3L is an important cofactor for Dnmt3a (Chedin et al. 2002). Dnmt3L may additionally be involved in retrotransposon silencing during premeiotic genome scanning in male germ cells (Bourc'his and Bestor 2004), since deletion of *Dnmt3L* in early male germ cells prevents de novo methylation of dispersed retrotransposons and causes meiotic failure in spermatocytes.

## DNMT METHYLTRANSFERASE ACTIVITY: A COMPLEX ISSUE

DNMTs have commonly been classified as either "maintenance" (DNMT1) or "de novo" (DNMT3) methyltransferases. This classification is based on the observation that Dnmt1 interacts with proliferating-cell nuclear antigen (PCNA) (Chuang et al. 1997), an auxiliary component of the DNA replication complex, and localizes to replication foci (Leonhardt et al. 1992). Yet it is emerging with increasing clarity that this classification is far too simplistic.

In human colorectal cancer cells, for example, there is evidence that DNA methylation patterns are maintained not by DNMT1 alone but by cooperation between DNMT1 and DNMT3B (Rhee et al. 2000, 2002; Ting et al. 2004). The effects of *Dnmt3a* and *Dnmt3b* disruption in ES cells likewise indicate that both Dnmt3a and Dnmt3b are involved in maintaining DNA methylation patterns (Chen et al. 2003). Dnmt1, on the other hand, shows little or no de novo methylation activity in vivo. Li and coworkers have recently proposed a model for the action of these three DNMTs (Chen et al. 2003): DNMT1 would be the main maintenance enzyme, acting with high efficiency but not full accuracy.

DNMT3A and DNMT3B, via their de novo activity, would act as "proofreaders," restoring CpG methylation at sites left untouched by DNMT1.

DNMT3L shows no methyltransferase activity, but it is nevertheless involved in the regulation of DNA methylation. As mentioned above, it contributes particularly to establishing genomic imprinting during gametogene-sis. It would appear to act as a cofactor for Dnmt3a, enhancing the latter's de novo activity (Bourc'his et al. 2001; Bourc'his and Bestor 2004; Kaneda et al. 2004).

Although DNMT2, as mentioned above, has retained only one of the domains characteristic of DNMTs, the methyltransferase domain, it was not shown until recently to be catalytically active (Hermann et al. 2003). It was also shown to display a certain sequence specificity for centromeric structures. This recent observation will likely revive interest in this still-mysterious member of the DNMT family.

## How Do DNMTs Interfere with Transcription?

DNMTs participate in gene silencing, but how? It has been known for many years that DNA methylation and chromatin structure are connected. In mammalian genomes, for example, high levels of DNA methylation coincide with heterochromatic regions (Razin and Cedar 1977). Also, methylated CpG islands (such as those of the female-inactivated X chromosome) appear in closed, transcriptionally silent chromatin with deacetylated histones, whereas unmethylated islands in gene promoters are transcriptionally favorable and have an open chromatin structure with highly acetylated histones (Bird and Wolffe 1999).

The mechanistic basis ofthelink between DNA methylation andchromatin structure has long remained obscure, but the recent explosion in knowledge on how chromatin organization modulates gene transcription has paved the way towards elucidatingthis link.

## CROSS-TALK AND TRANSCRIPTIONAL SILENCING

Initial papers from the laboratories of A. Bird and A. Wolffe were the first to unveil a mechanistic connection between DNA methylation and histone modification. They showed that methyl-CpG binding domain

(MBD) proteins, which selectively recognize methylated CpG dinucleotides, are components of—or establish contacts with—histone deacetylase (HDAC) complexes (Jones et al. 1998; Nan et al. 1998). HDACs remove acetyl groups from histone tails and help to maintain nucleosomes in a compact, transcriptionally silent state.

Next, a much more direct connection between CpG methylation and deacetylation was identified: DNMTs appear to repress transcription through recruitment of histone deacetylases (Burgers et al. 2002). The fact that each DNMT associates with HDAC prompts the question: Why is this contact necessary? One clue might lie in the ability of DNMTs to act as maintenance and/or de novo methyltransferases. A challenge for the cell is to restore in newly replicated DNA the chromatin structure needed to maintain the tran-scriptional activity states dictated by chromatin modifications. In the case of at least one maintenance enzyme, DNMT1, its association with HDAC is particularly attractive: It occurs predominantly at replication foci during the late S-phase, when most of the heterochromatin is duplicated (Rountree et al. 2000). DNMT1 may thus be necessary to ensure that the histones forming the nucleosomes assembled at newly replicated sites are deacetylated.

An unexpected finding has emerged from the study of the DNMT–HDAC interaction, mediated by the non-catalytic N-terminal portion of the DNMT. Intriguingly, transcriptionalsilencing does not require preservation ofDNMT enzymatic activity. In addition, Dnmt3L can still recruit the HDAC repressive machinery despite its lack of DNMT activity (Deplus et al. 2002).

It thus seems that DNMTs can carry out some HDAC-associated functions independently of their ability to methylate CpG sites, at least in certain circumstances. Although these observations remain to be confirmed in vivo, it is tempting to speculate that DNMTs are more versatile than initially anticipated. In other words, they may be multifaceted proteins performing other functions in addition to methylation of CpG dinucleotides.

More recently, DNMTs have been implicated in another chromatin-related transcriptional repression process, involving methylation of histone H3 at lysine 9. This connection was first evidenced in the ascomycete fungus *Neu-rospora crassa* and in the plant *Arabidopsis thaliana* by E. Selker's and S. Jacobsen's groups, respectively (Jackson et al. 2002; Selker et al. 2003). It was shown that mutations in the genes *dim-5* of Neurospora and *kryptonite* of Ara-bidopsis result in loss of DNA methylation in these organisms. Excitement arose from the finding that these genes encode H3-K9 histone methyltrans-ferases.

The mechanisms linking DNA methylation to histone methylation remain unclear and there are likely several ways that connect these two epigenetic events. In Arabidopsis, the adaptor protein LHP1 (the homolog of the mammalian heterochromatin protein 1, HP1) is not needed to maintain DNA methylation, and at least deacetylase HDA6 is instead required (Bender 2004). In Neurospora, however, the HP1 protein could be a possible link between DNA and histone methylation, since it has been shown that HP1 is required for DNA methylation (Selker et al. 2003). According to the current working model in Neurospora, methylation at H3-K9 by DIM5 would create a binding platform for HP1. This adaptor protein would then recruit the DIM2 DNMT. In this way, histone methylation would influence DNA methylation.

Might sucha model also apply to mammals? Is there cross-talk in mammals between histones and DNA and are DNMTs involved? Recent studies indicate that this maywell be the case. DNMTs appear to associate with histone methyl-transferase activities that modify lysine 9 of H3. Interaction with the Suv39h histone methyltransferase may be involved (Fuks et al. 2003a; Lehnertz et al. 2003). Contact between DNMTs and proteins HP1á and HP1â has also been demonstrated (Fuks et al. 2003a; Lehnertz et al. 2003). Lastly, results obtained with Suv39h-double-null mouse embryonic stem cells indicate that Suv39h-mediated H3-K9 trimethylation can direct Dnmt3b to major satellite repeats present in pericentric heterochromatin (Lehnertz et al. 2003). As in the proposed Neurospora model, mammalian DNMTs might thus interact with the adaptor molecule HP1 and be present in the vicinity of chromatin-containing methylated histones, this leading to CpG methylation and gene silencing.

Because mammals possess several H3-K9 methyltransferases, the methy-lation "conversation" is likely to be more complex in mammals than in Neu-rospora. This view is supported by work on Suv39h-double-null cells, based on the use of highly specific antibodies that discriminate between H3-K9 di- and trimethylation. In this study, pericentric major satellites displayed Suv39h-dependent H3-K9 trimethylation, while centromeric minor satellites showed a "preference" for Suv39h-independent H3-K9 dimethylation (Lehn-ertz et al. 2003). As Dnmt3b-dependent DNA methylation at minor satellites was unimpaired in Suv39h-double-null cells, it could be that H3-K9 dimethylation catalyzed by some other enzyme, the identity of which is still unknown, is responsible for the observed targeting of Dnmt3b. Other H3-K9 methyl-transferases, such as G9a (Xin et al. 2003) or SETDB1 (Ayyanathan et al. 2003) are reported to regulate DNA methylation-mediated gene silencing, but whether these enzymes associate directly with DNMTs to affect CpG methy-lation remains to be seen.

## Histone and DNA Methylation: Mutual Boosting and Feedback Loops

The observations just described suggest a straight line from H3-K9 methyla-tion to DNA methylation. This is consistent with evidence showing that only genes silenced by other mechanisms are subject to CpG methylation, which would thus be a secondary event in gene silencing (Bird 2002). Recent data indicate that DNA methylation might in turn exert a feedback effect on ly-sine methylation, this leading to mutual reinforcement of these two distinct methylation layers.

Work on MBD proteins supports this view. At the site of a methylated gene that it regulates, MeCP2 was found to facilitate methylation of histone H3 at lysine 9, likely catalyzed by the Suv39h enzyme (Fuks et al. 2003b; our unpublished data). In addition, methylation by the H3-K9 enzyme SETDB1 was shown to depend on MBD1 and on DNA methylation at specific loci (Sarraf and Stancheva 2004).

Whether DNMTs dictate histone methylation directly has not been reported to date, but work on Arabidopsis may again lead the way. Mutational analyses indicate that MET1, the plant homologof mammalian DNMT1, influences H3-K9 methylation (Soppe et al. 2002). Thus, while much more work is needed to extend these studies to other settings, it would seem that epigenetic information, embodied in residue methylation states, can flow from histones to DNA and back. This would be similar to the established flow of information between deacetylated histones and methylated DNA, involving physical association of MBDs and DNMTs with HDACs and resulting in feedback loops (Jaenisch and Bird 2003).

All this suggests that DNA methylation may lead to gene silencing as part of an epigenetic program carried out through the interactions illustrated in Fig. 2. In the initial phase, DNMTs bound to an adaptor molecule such as HP1 would add methyl groups to DNA only on chromatin that is methylated at lysine 9 of histone H3. Association of the DNMTs with an H3-H9 methyltrans-ferase (e.g., Suv39h) would ensure a direct impact ofH3-K9 methylation states on the DNMTs. These would also make contacts with HDACs. This would lead to partial gene silencing. In a second step, the generation of methylated DNA by the DNMTs would permit binding of MBDs to DNA. The bound MBDs would in turn interact with H3-K9 methyltransferase and facilitate lysine methylation. As deacetylation of histone H3 at lysine 9 is necessary for methy-lation to take place on this residue (Rea et al. 2000), deacetylation of histone H3 at lysine 9 would be followed by histone methylation, which in turn might result in the recruitment of proteins such as HP1.

It will be essential in the future to unravel in more detail the precise sequence of events. Multiple mechanisms are likely to contribute to the establishment and maintenance of silenced epigenetic states. Nevertheless, the above model is attractive because it suggests that DNA methylation might act together with histone deacetylation and H3-K9 methylation to generate a self-reinforcing cycle and thereby perpetuate and maintain a repressed chromatin state.

## DNMT AND HDAC CONNECTION

### How Are DNMTs Targeted to Precise DNA Sequences?

Methylated cytosines are not randomly distributed in the mammalian genome. The mechanisms underlying the establishment of DNA methylation patterns remain largely a mystery. Methylation patterns are generated by the DNMTs, and evidence is accruing that DNMTs have preferred sites of action. Targeted disruption of Dnmt3a and Dnmt3b in mouse embryonic stem cells has demonstrated that they have some overlapping sites, while each also has its specific targets. For example Dnmt3b, but not Dnmt3a, participates in the methylation of centromeric minor satellite repeats (Okano et al. 1999). Likewise, studies on DNMT3B mutations causing a rare human condition called ICF (for immunodeficiency, centromere instability, and facial anomalies) suggest that DNMT3B methylates specific centromeric repeats (Xu et al. 1999). Experiments using a stable episomal system also show that Dnmt3a and Dnmt3b may have some distinct preferred target sites (Hsieh 1999).

How DNMT activity is preferentially targeted to specific regions of the genome is still poorly understood. DNMTs do not appear to have an intrinsic capacity to discriminate among primary nucleotide sequences.

## Chromatin-Based Targeting

One possibility is that chromatin-modifying or-remodeling proteins might be required to attract DNMTs to DNA to be methylated. As illustrated in the previous section, emerging clues suggest that de novo DNMTs take cues from histone modifications. On the one hand, methylation at lysine 9 of H3 can facilitate CpG methylation, and DNMTs associate with H3-K9 enzymatic activity. On the other hand, DNMTs interact directly with histone deacetylases. In Neurospora, HDAC inhibition by trichostatin A (TSA) causes specific cyto-sine hypomethylation (Selker 1998). Moreover, transient transfection studies suggest that histone acetylation may dictate, in some instances, DNA methy-lation (Cervoni and Szyf 2001). The current model proposes that DNMTs might be targeted to a genomic sequence by nucleosomes featuring histone hypoacetylation or H3-K9 methylation. Thus, histone modifications would provide a basis for the generation of CpG methylation patterns by DNMTs.

In addition to histone modification, chromatin remodeling might be required for DNMT-catalyzed methylation. Emerging clues point to the possibility that chromatin remodeling might be needed to give DNMTs access to chromatin templates that would otherwise remain inaccessible. Studies on Arabidopsis, mice, and humans indicate that loss or alteration of DNA methy-lation may result from mutations in

SNF2-like ATPases or from disruption of the corresponding genes (Meehan and Stancheva 2001), i.e., chromatin-remodeling proteins requiring ATP in order to disrupt histone–DNA interactions and to enable nucleosomes to slide along the DNA.

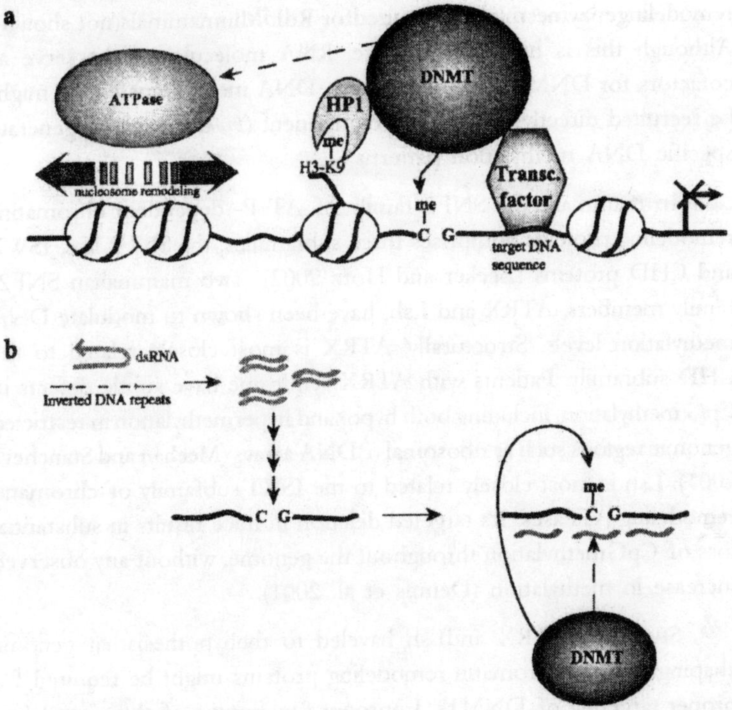


Fig.a, b Possible mechanisms for the targeting of DNMTs to specific DNA sequences. a Chromatin- and transcription factor-based targeting. Histone methylation at Lys 9 of H3 influences DNA methylation, possibly through recruitment of DNMTs by the adaptor HP1. HDACs associate with DNMTs (not shown) and may also provide a basis for the generationofCpGmethylation patternsbyDNMTs.ATP-dependentchromatin-remodeling proteins such as Lsh or ATRX could recruit DNMTs and, although this has yet to be demonstrated, might directly assist methylation of CpGs by DNMTs (*broken arrow*). Targeting of DNMTs may also be achieved through their association with specific transcription factors such as PML/RAR or Myc, with subsequent CpG methylation in the targeted promoter. b Do DNMTs "listen" to RNA?

It seems that RNA-mediated DNA methylation (RdDM) can occur in mammals (Morris et al. 2004; Kawasaki and Taira 2004).Double-stranded (ds)RNA is processedbythe Dicer enzyme into small interfering (si)RNAs. By analogy to what happens in plants, chromatin-modifyingand/or-remodelingenzymesmightberequiredfor RdDMinmammals(not shown). Although this is highly speculative, RNA molecules might serve as cofactors for DNMTs. In other words, DNA methyltransferases might be recruited directly by an RNA component (*broken arrow*) to generate specific DNA methylation patterns

In mammals, the SNF2 family of AT P- dependent chromatin-remodeling proteins comprises three subfamilies: the SNF2-like, ISWI, and CHD proteins (Becker and Horz 2002). Two mammalian SNF2-family members, ATRX and Lsh, have been shown to modulate DNA methylation levels. Structurally, ATRX is most closely related to the CHD subfamily. Patients with ATRX syndrome have subtle defects in CpG methylation, including both hypo- and hypermethylation in restricted genomic regions such as ribosomal (r)DNA arrays (Meehan and Stancheva 2001). Lsh is most closely related to the ISWI subfamily of chromatin remodeling ATPases. Its targeted deletion in mice results in substantial loss of CpGmethylation throughout the genome, without any observed increase in methylation (Dennis et al. 2001).

Studies onATRX andLsh haveled to thehypothesisthat genome shapingby these chromatin-remodeling proteins might be required for proper targeting of DNMTs. Improper functioning of the remodeling enzymes may lead to either hypo- or hypermethylation (as observed in ATRX patients). The latter effect may be due to aberrant targeting of DNMTs to regions that would not normally be methylated.

What could be the mechanisms by which the SNF2 ATPases alter methy-lation patterns? Are remodeling proteins directly contacting DNMTs to regulate their chromatin accessibility? Recent biochemical studies on DNMT3B may point to a direct connection between DNMTs and remodeling enzymes. Endogenous DNMT3B was found to associate with ATPase activity and to interact in vivo with the ATP-dependent chromatin-remodeling enzyme hSNF2H (Geiman et al. 2004). It will be crucial in the future to evaluate whether hSNF2H modulates CpG methylation patterns as observed for ATRX and Lsh.

When envisaging a potential direct link between DNMTs and SNF2 AT-Pases, it is necessaryto consider a number of additional issues. One question is whether remodeling enzymes influence de novo DNMT activity, maintenance DNMT activity, or both. For example, work on Lsh indicates that synthesis of this protein correlates with the S-phase of the cell cycle. It has been postulated that Lsh might facilitate access of DNMTs to hemimethylated sites after replication occurs and thus contribute to maintaining methylation patterns (Dennis et al. 2001).

Another question that researchers are eager to answer is whether ATP-dependent nucleosome-remodeling enzymes can directly assist methylation of CpG residues by DNMTs. A fruitful approach might be to develop in vitro assays employing recombinant SNF2 ATPases and DNMTs with reconstituted chromatin substrates.

## TARGETING OF DNMTS BY DNA-BOUND TRANSCRIPTION FACTORS

Another possible mechanism for the recruitment of DNMTs to specific genome sequences might involve their association with specific transcription factors. Early work did point in this direction: It was found that DNA-binding transcriptional repressors such as E2F or RP58 can recruit DNMTs to their target promoters and thereby cause transcriptional repression (Burgers et al. 2002). Disappointingly, however, this repression was found not to depend on the methyltransferase activity of the DNMTs.

A breakthrough came from studies focusing on another transcriptional regulator, PML-RAR. This oncogenic protein, generated by a translocation, appears in acute promyelocytic leukemia. It was found that PML-RAR can recruit DNMT1 and DNMT3A to the retinoid acid receptor (RAR)â promoter, this leading to hypermethylation of the promoter and to gene silencing (Di Croce et al. 2002). This was the first demonstration that DNMTs can be recruited by a DNA-bound transcriptional repressor, with subsequent CpG methylation of the targeted-promoter.

It is tempting to draw a parallel between the targeting of DNMTs to promoters by specific DNA-binding proteins and the mechanisms by which chromatin-modifying enzymes regulate gene expression by establishing local changes in chromatin structure. For instance, histone-modifying enzymes such as acetylases and deacetylases are targeted to promoters via their association with DNA-bound activators or repressors, and this appears as a general strategy for delivering the corresponding enzymatic activities to specific promoters (Kurdistani and Grunstein 2003). By analogy, cells might use a similar general strategy to target DNMTs to precise loci.

How general a mechanism is DNMT targeting by transcription factors? Recent work in our laboratory shows that the Myc transcription factor associates in vivo with Dnmt3a and targets its enzymatic activity—through the DNA-binding protein Miz-1—to the p21Cip1 promoter. In this system, DNA methylation is required for Myc-mediated repression of p21Cip1 (Brenner et al. 2005). What's more, yeast-two hybrid screens using DNMTs as baits have led to the identification of known transcription factors that could potentially target their activity to specific promoters (our unpublished data). Thus, it is reasonable to hypothesize that DNMT-catalyzed CpG methylation steered by sequence-specific binding proteins may be a general mechanism for the establishment of DNA methylation patterns.

## The RNA Trigger

Another potential mechanism for the establishment of DNA methylation patterns in mammals could involve RNA. This exciting possibility is attracting more and more attention. It is known that in plants, post-transcriptional gene silencing—which resembles RNA interference (RNAi)—triggers DNA methylation. RNAi is activated by the expression of dsRNA, which provides a trigger for the degradation of transcripts with which it shares sequence identity. siRNAs 21–26 nucleotides in length are key actors in RNAi, deriving from dsRNA through the action of the RNase III Dicer enzyme (Matzke and Birchler 2005).

Promoter sequence-containing dsRNA can cause gene silencing by DNA methylation of the homologous promoter regions. This RNA-directed DNA methylation (RdDM) is highly sequence-specific and largely confined to regions of RNA–DNA sequence homology (Matzke and Birchler 2005). With the help of molecular genetics, investigators are beginning to unravel the mechanisms underlying RdDM in plants. These studies reveal that RdDM requires various proteins: RNAi-pathway proteins, a novel remodeling enzyme, and also histone-modifying enzymes and DNMTs. Although RdDM seems to be a common and general mechanism for silencing gene transcription in plants, this is likely not the case in Neurospora. Notably, DNA methylation occurs normally in the latter organism in the absence of key elements of the RNA-silencing machinery (Freitag et al. 2004).

Clearly, RdDM is not a general DNA methylation-targeting mechanism. This prompts several questions: Does RNA-directed DNA methylation mechanism exist in mammals? If so, does it involve DNMTs and what is their role? It has long been known that in mammals, non-coding RNAs are involved in processes such as allelic imprinting and X inactivation. For instance, studies on mice have shown that expression of *Xist*, a non-coding RNA involved in X inactivation, is regulated by expression of its antisense RNA *Ts i x*, driven by a promoter downstream from the *Xist* gene (Lee and Lu 1999). Also, an RNA component is required to maintain the structure of mouse pericentromeric heterochromatin (Maison et al. 2002). Furthermore, studies focusing on rearrangement of the á-globin gene in a patient with á-thalassemia showed that the á-globin gene on the rearranged chromosome was intact but silenced epi-genetically through convergent transcripts correlating with DNA methylation (Tu farelli et al. 2003). As yet, however, there is no clear evidence that these chromatin-based regulations involve RNA-directed silencing.

More recent work has yielded a confused picture regardingthe involvement of RNA-mediated CpG methylation in mammals. Studies on mouse oocytes suggest that dsRNA expression, while inducing post-transcriptional silencing by RNAi, does not induce sequence-specific methylation of the cognate DNA sequence (Svoboda et al. 2004). Limitations to this study were that the system usedwas confinedto a specific celltype andthat RdDM targeting was analyzed in a single intronless endogenous gene. Two other reports suggest, on the contrary, that RNA-mediated DNA methylation can occur in mammals. In one study on human kidney cells, siRNA targeted to a promoter by means of lentiviral transduction was found to silence the endogenous EF1A gene, silencing being associated with DNA methylation (Morris et al. 2004). In another work, synthetic siRNAs targeted to the E-cadherin gene in human breast epithelial cells caused its transcriptional repression (Kawasaki and Taira 2004).

Studies in which expression of DNMT genes was suppressed by means of siRNAs targeting the corresponding messenger (m)RNAs have shown that DNMT1 and DNMT3B, but not DNMT2, are likely necessary for siRNA-mediated transcriptional silencing of expression from the E-cadherin promoter. Bisulfite sequencing revealed a correlation between E-cadherin silencing correlates and sequence-specific CpG methylation (Kawasaki and Taira 2004). Thus, RdDM appears also to occur in mammals. Yet from the few reports available to date, it would already seem that induction of DNA methylation by siRNA in mammalian cells is not a general phenomenon.

If it turns out to occur in mammals in a limited range of situations, it will be important to determine which situations, and to explain why only some cells or some genes are susceptible to RdDM. It will also be essential to unravel the underlying mechanisms. Key questions will be: How are siRNAs guided to genomic DNA? How do they gain access to it? Also worthy of special attention, given the mechanism of RdDM in plants, will be the role played by chromatin-modifying and-remodeling enzymes and the sequence of events leading to siRNA-directed DNA methylation.

Regarding DNMTs, it will be important to determine how they are mechanistically connected to the RNAi machinery. While these are still early days, one might imagine, for instance, that RNA molecules serve as cofactors for DNMTs, thereby guiding CpG methylation to precise sequences.

The recent observation that DNMT3A and DNMT3B can interact, at least in vitro, with RNA molecules is intriguing (Jeffery and Nakielny 2004). Hence, although highly speculative, the possibility that DNMTs might be targeted directly by an RNA component to establish specific DNA methylation patterns may deserve future study.

## Conclusions

Since the isolation and characterization of the DNMTs in the 1990s, abundant evidence hasestablished their roleas keyregulatorsof DNA methylation. What is changing is our idea of how DNMTs cause transcriptional repression and our understanding of how chromatin structure is regulated. It seems almost certain that chromatin modifications and DNMTs are tightly linked in mammals. As discussed here, clues are emergingthat DNMTs mayact together with histone deacetylation and H3-K9 methylation to generate a self-reinforcing cycle that perpetuates and maintains a repressed chromatin state. Despite rapid growth of knowledge on the intimate link between chromatin and DNMTs, t he picture is still blurred. It will be a notable challenge to untangle the mutualreinforcementsof repression andthe different statesof chromatin- and DNA-modifying activities required to silence different genomic regions (e.g., highly repetitive elements versus single-copy genes). What's more, the observation that DNMTs may also silence gene expression by recruiting histone deacetylase and H3-K9 methyltransferase rather than through their ability to methylate CpG sites had led to the tempting speculation that DNMTs might be multifaceted proteins with broader roles in transcriptional repression than first anticipated.

The origin of DNA methylation patterns is a longstanding mystery. Recent studies are providing clues that may help explain how DNMTs are targeted to preferred genomic loci. Like chromatin-modifying enzymes (e.g., HDAC), DNMTs are recruited to promoters by repressors of transcription, this leading to gene silencing. We anticipate a flurry of research aiming to identify transcription factors capable of targeting DNMTs to specific genes. If this mechanism of DNMT targeting turns out to be general, a key issue will be to understand precisely how specificity is achieved with respect to the DNMT-recruiting transcription factor.

Finally, exciting new evidence suggests a connection between RNAi-mediated pathways and DNA methylation in mammals. Whether DNMTs "listen" directly to RNA remains an open question. Work shedding light on this question is eagerly awaited.

# 4

# DNA Methylation in Plants

DNA in plants is highly methylated, containing 5-methylcytosine ($m^5C$) and $N^6$-methyladenine ($m^6A$); $m^5C$ is located mainly in symmetrical CG and CNG sequences but it may occur also in other non-symmetrical contexts. $m^6A$ but not $m^5C$ was found in plant mitochondrial DNA. DNA methylation in plants is species-, tissue-, organelle- and age-specific. It is controlled by phytohormones and changes on seed germination, flowering and under the influence of various pathogens (viral, bacterial, fungal). DNA methylation controls plant growth and development, with particular involvement in regulation of gene expression and DNA replication. DNA replication is accompanied by the appearance of under-methylated, newly formed DNA strands including Okazaki fragments; asymmetry of strand DNA methylation disappears until the end of the cell cycle.

A model for regulation of DNA replication by methylation is suggested. Cytosine DNA methylation in plants is more rich and diverse compared with animals. It is carried out by the families of specific enzymes that belong to at least three classes of DNA methyltransferases. Open reading frames (ORF) for adenine DNA methyltransferases are found in plant and animal genomes, and a first eukaryotic (plant) adenine DNA methyltransferase *(wadmtase)* is described; the enzyme seems to be involved in regulation of the mitochondria replication. Like in animals, DNA methylation in plants is closely associated with histone modifications and it affects binding of specific proteins to DNA and formation of respective transcription complexes in chromatin.

The same gene (*DRM2*) in *Arabidopsis thaliana* is methylated both at cytosine and adenine residues; thus, at least two different, and probably interdependent, systems of DNA modification are present in plants. Plants seem to have a restriction-modification (R-M) system. RNA-directed DNA methylation has been observed in plants; it involves de novo methylation of almost all cytosine residues in a region of siRNA-DNA sequence identity; therefore, it is mainly associated with CNG and non-symmetrical methylations (rare in animals) in coding and promoter regions of silenced genes. Cytoplasmic viral RNA can affect methylation of homologous nuclear sequences and it may be one ofthe feedbackmechanisms between the cytoplasm and the nucleus to control gene expression.

## INTRODUCTION

DNA in plants is highly methylated, containing additional methylated bases such as 5-methylcytosine ($m^5C$) and*N*$^6$-methyladenine ($m^6A$). DNA methyla-tion in plants is species-, tissue-, organelle- and age-specific. Specific changes in DNA methylation accompany the entire life of a plant, starting from seed germination up to the death programmed or induced by various agents and factors of biological or abiotic nature. In fact, the ontogenesis and the life itself are impossible without DNA methylation, because this genome modification in plants, like in other eukaryotes, is involved in a contiol of all genetic functions including transcription, replication, DNA repair, gene transposition and cell differentiation. DNA methylation controls plant growth and development. On the other hand, plant growth and development are regulated by specific phytohormones, and modulation of DNA methylation is one of the modes of the hormonal action in plant.

Plant DNA methylation has many things in common with it in animals but it has also specific features and even surprises. Plants have a more complicated system of genome methylations compared with animals; besides, unlike animals, they have the plastids with their own unique DNA modification system that may control plastid differentiation and functioning; DNA methylation in plant mitochondria is performed in a different fashion compared with nuclei.

Plants seem to have a restriction-modification (R-M) system. Plants supply us with unique systems or models of living organisms that help us to understand and decipher the intimate mechanisms and the functional role of genome modification and functioning in eukaryotes.

Some features and regularities of DNA methylation in plants are described in this chapter, which cannot be a comprehensive elucidation of many complicated problems associated with this genome modification in the plant kingdom. An interested reader may find the intriguing details of plant DNA methylation and its biological consequences also in available reviews (Fedo-roff 1995; Meyer 1995; Richards 1997; Dennis et al. 1998; Finnegan et al. 1998b; Colot and Rossignol 1999; Kooter et al. 1999; Finnegan et al. 2000; Finnegan and Kovac 2000; Matzke et al. 2000; Sheldon et al. 2000; Wassenegger 2000; Bender 2001; Chaudhury et al. 2001; Martienssen and Colot 2001; Paszkowski and Whitham 2001; Vaucheret and Fagard 2001; Bourc'his and Bestor 2002; Kakutani 2002; Li et al. 2002; Wassenegger 2002; Liu and We n del 2003; Stokes 2003; Vinkenoog et al. 2003; Matzke et al. 2004; Montgomery 2004; Scott and Spielman 2004; Steimer et al. 2004; Tariq and Paszkowski 2004).

## CYTOSINE DNA METHYLATION

### Chemical Specificity

5-Methylcytosine in plant DNA is mainly located in symmetrical CG and CNG sequences (Gruenbaum et al. 1981; Kirnos et al. 1981; Kovarik et al. 1997), but it is found also in various non-symmetrical sequences (Meyer et al. 1994; Oakeley and Jost 1996; Goubely et al. 1999; Pelissier et al. 1999). Some plant cytosine DNA methyltransferases may methylate any cytosine residue in DNA except for in CpG, and the specificity of the enzyme is mainly limited by the availability of certain cytosines in the chromatin structure that can be modulated essentiallybythe enzyme itself or its complexeswith other proteins (Wada et al. 2003). The share of $m^5C$ located in CNG sequences in plant DNA may correspond to up to about 30% of total $m^5C$ content in the genome (Kirnos et al. 1981).

The finding of $m^5C$ in these sequences in plant DNA was the first safe and widely accepted evidence of the non-CG methylation in eukaryotes. For a long period many investigators involved in the DNA methylation research were very sceptical about the existence of this type of DNA methylation in animals, despite the respective obvious data that were alreadyavailable (Salomon andKaye1970; Sneider 1972; Woodcocket al. 1987; Toth et al. 1990; Clark et al. 1995). The non-CG methylation is carried out by the Dnmt3a/Dnmt3b enzyme(s) in mammalian cells (Ramsahoye et al. 2000) and dDnmt2 in *Drosophila* cells (Lyko 2001) and seems to be guided by RNA (Matzke et al. 2004).

It should be mentioned that attention to the significance of this particular DNA methylation type for proper genome functioning in animal cells is still underpaid, and in some modern epigenomic projects even neglected. But this particular genome modification in animals seems to have a physiological sense. For example, the histone deacetylase inhibitor valproate increased 5-lipoxygenase the messenger (m)RNA level and reduced CNG methylation of the 5-lipoxygenase core promoter in human neuronlike NT2-N but not in NT2 cells (Zhang et al. 2004). The situation with CNG (non-CG methylation) in plants is better because this modification is definitely involved in the epigenetic gene silencing including small interfering (si)RNA-directed silencing (Bartee et al. 2001; Bender 2001; Lindroth et al. 2001).

## BIOLOGICAL SPECIFICITY

### Species Specificity

Very high $m^5C$ content (up to about 9 mol%) in total DNA is a specific feature of plants (Vanyushin and Belozersky 1959); in some cases in plant *(Scilla sibir-ica)* satellite DNA, the cytosine moiety is almost completely represented by $m^5$ C. In earlier days, we even could not rule out the possibility that $m^5$ C might be incorporated into plant DNA in a ready-made form at the template level during DNA synthesis; there is an indication that 5-methyl-2'-deoxycytidine 5'-triphosphate may be incorporated into DNA in animal cells (Nyce 1991). But none of any methyl-labelled $m^5$ C derivatives was found to be incorporated into DNA in an intact plant, and it was concluded that all $m^5C$ present in plant DNA is a product of DNA methylation (Sulimova et al. 1978). Thus, DNA in plants compared with other organisms is the most heavily methylated. $m^5C$ was found in DNA of all archegoniate (mosses, ferns, gymnosperms and others) and flowering plants (dicots, monocots) investigated. As a rule, DNA of gymnosperm plants contain less $m^5C$

than DNA of flowering plants (Vanyushin and Belozersky 1959; Vanyushin et al. 1971). The species differences of phylogenetic significance in the frequency of methylated CNG sequences in genomes of plants are clearly pronounced (Kovarik et al. 1997; Fulnecek et al. 2002).

## Age Specificity

It was known that aging in animals is accompanied by a global DNA demethy-lation, with the amount of $m^5C$ in the DNA of all organs essentially decreased (Vanyushin et al. 1973). A similar situation takes place in plants: The amount of $m^5C$ decreases and its distribution among pyrimidine isopliths in DNA is essentially changed on seed germination (Sulimova et al. 1978). Some DNA sequences unmethylated in seeds become methylated in seedlings. The age changes in DNA methylation may have a regulatory character and seem to be associated with a developmental switch-over of the gene functioning (Sulimova et al. 1978). Age differences in the DNA methylation patterns were found in various plants (Fraga et al. 2002;Baurens et al. 2004; Xionget al.1999).

## Cellular (Tissue) Specificity

Similarly to animal DNA (Vanyushin et al. 1970), the $m^5C$ content in plant DNA is tissue (cellular) specific (Vanyushin et al. 1979). This may reflect an association of DNA methylation with cellular differentiation in plants. There are many data available now indicating that methylation patterns oftotalDNA and distinct genes in various tissues of the same plant are different (Bianchi and Viotti 1988; Lo Schiavo et al. 1989; Riggs andChrispeels 1999; Palmgren et al. 1991; Kutueva et al. 1996; Rossi et al. 1997; Ashapkin et al. 2002; Chopra et al. 2003). The $m^5C$ content in DNA from different plant tissues is associated with a flowering gradient: It is higher in generative tissues of pea, tobacco, apple tree and lily-of-the-valley plants compared with vegetative tissues (Chvojka et al. 1978).

The gene silencing associated with DNA methylation is tissue specific also; methylation of a glucuronidase reporter gene in the transgenic rice plant accompanied by loss of expression was initially restricted to the promoter region and observed in the vascular bundle tissue only, the expression character was similar to that of a promoter with a deleted vascular bundle expression element (Klöti et al. 2002).

## Subcellular (Organelle) Specificity

In plant cells the nuclear, mitochondrial and plastid DNAs are methylated in a different fashion. Contrary to animals (Vanyushin and Kirnos 1974), in plants m$^5$C was not found in mitochondrial (mt)DNA (Aleksandrushkina et al. 1990). Instead, plant mtDNA does contain m$^6$A, with about 0.5% adenine in mtDNA from wheat seedlings being methylated (Vanyushin et al. 1988). DNA of plastids (chromoplasts, leucoplasts, amyloplasts) contains various methylated bases including m$^5$ C and m$^6$ A, but the chloroplast DNA practically is not methylated (Ngernprasirtsiri et al. 1988; Ngernprasirtsiri and Akazawa 1990; Fojtova et al. 2001). It was assumed that plastid DNA (de)methylation is associated with differentiation of plastids and, in particular, with photosynthetic gene functioning in chloroplasts (Ngernprasirtsiri and Akazawa 1990).

## Intragenome Specificity

Plant nuclear DNA is unevenly methylated, since m$^5$C is mainly located in GC-enriched and highly repetitive sequences (Guseinovet al. 1975; Guseinov and Vanyushin 1975). In particular, in petunia the repetitive DNA sequences (RPS) have hot spots for de novo DNA methylation; for example, the palin-dromic, moderately to highly RPS-repetitive element that is not predominantly localized to constitutive heterochromatin is a target for strong de novo methylation. It seems to be due to an intrinsic signal formed by unique DNA secondary structure. This palindromic element, integrated into the genome of *Arabidopsis thaliana*—a plant lacking homology to the RPS—acts as a de novo hypermethylation site in the non-repetitive genomic background of *Arabidopsis*, strongly suggesting a signal function of the palindromic RPS unit (Muller et al. 2002).

The bulk of the repetitive DNA constitutes transposons and retrotrans-posons; the repeats are primary targets of methylation both in flowering (ten Lonuis et al. 1995a, b; Muller et al. 2002) and archegoniate *(Marchantia paleacea)* (Fukuda et al. 2004) plants. Although the repetitive elements are methylated in both plants and animals, most mammalian exons are methylated but plant exons are mainly not; there is even an opinion that targeting of methylation specifically to transposons is restricted to plants (Rabinowicz et al. 2003).

Usually retrotransposons are hypermethylated (Fukuda et al. 2004) and their transcription is activated by demethylation (Komatsu et al. 2003). Silent retrotransposons can be reactivated by *ddm1* mutation (Hirochikaet al. 2000). In accordance with methylation patterns, the maize transposable activator (Ac) elements were divided into two distinct groups. About 50% of the elements are fully unmethylated at cytosine residues through the 256 nucleotides at the 5'-end (the promoter region), whereas the other half is partially methylatedbetween Ac residues 27 and 92. In contrast, at the3'-end,all Ac are heavily methylated between residues 4372 and 4554; the more internally located Ac sequences and the flanking *waxy* DNA are unmethylated. Methylated cytosines in Ac are located in both the symmetrical (CG, CNG) and non-symmetrical sequences (Wang et al. 1996).

Complex cereal genomes are largely composed of small gene-rich regions intermixed with 5- to 200-kb blocks of repetitive DNA. The repetitive DNA blocks are usually methylated at 5'-CG-3' and 5'-CNG-3' cytosines in most or all adult tissues, while the genes are generally unmethylated at these sites (Yuan et al. 2002).

The activity and inactivity of endogenous retrotransposon Tos17 in calli and regenerated rice plants are accompanied by hypo- and hyper-CG methylation and hemi and full CNG methylation, respectively, within the element, whereas immobilization of the element in the other two plant lines is concomitant with near-constant, full hypermethylation. Treatment with 5-azacytidine (azaCyt) induced both CG and CNG partialhypomethylation of Tos17 in two lines (Matsumae and RZ35). A heritable alteration in cytosine methylation patterns occurred in three of seven genomic regions flanking Tos17 in calli and regenerated plants of RZ35, but in none of the five regions flanking dormant Tos17 in the other two lines (Liu et al. 2004).

In *Arabidopsis*, m$^5$C appears to be differentially distributed along the major ribosomal (r)RNA gene repeat. The most striking variation in cytosine methylation in the long arrays of rRNA genes was found at the tips of chromosomes 2 and 4 (Raddle and Richards 2002).

In *Brassica napus,* S1Bn short interspersed element (SINE) retroposons are twofold more methylated than the average methylation level of the nuclear DNA; S1Bn cytosines in symmetrical CG and CNG sites are methylated at a level of 87% and 44%, respectively; 5.3% of S1Bn cytosines in non-symmetrical positions were also methylated. Of this asymmetrical methylation, 57% occurred at a precise motif [Cp(A/T)pA] that only represented 12% of the asymmetrical sites in S1Bn sequences, suggesting that it represents a preferred asymmetrical methylation site. This motif is methylated in S1Bn elements at only half the level observed for the Cp(A/T)pG sites (Goubely et al. 1999).

The methylation patterns of various plant chromosomes are quite different, with even some regions of chromosomes showing enhanced or reduced methylation (Castilho et al. 1999); DNA in euchromatin is less methylated compared with heterochromatin DNA (Buzek et al. 1998; Fransz et al. 2002; Luchniak et al. 2002; Mathieu et al. 2002a). Heterochromatin in *Arabidopsis* determined by transposable elements and related tandem repeats is under the control of the chromatin remodelling ATPase DDM1 (Lippman et al. 2004).

The most methylated repeated family at CG, CNG and asymmetrical sites was found in the 5S ribosomal DNA. It was highly methylated (Fulnecek et al. 1998; Fulnecek et al. 2002) even though it is transcribed (Mathieu et al. 2002b). Thus, 5S rRNA gene expression is not inhibited by DNA methylation in *Arabidopsis* (Mathieu et al. 2002b).

As a rule, centromere regions and satellite plant DNA are heavily methylated with strand asymmetries (Luo and Preuss 2003). In *Viciafaba* metaphase chromosomes, the m$^5$C residues are present in different chromosomal sites and are particularly abundant in telomeric and/or subtelomeric regions and in certain intercalary bands (Frediani et al. 1996). In the *Melandrium album* male cells, a more intensive methylation on the shorter arm of the only X chromosome was observed in comparison with the longer X arm. A global hypermethylation of the male Y chromosome was not found. But in female cells, the specific cytosine methylation pattern of the X chromosome was found on a single X chromosome, whereas the other X displayed an overall higher level of $m^5C$ (Siroky et al. 1998).

At least two CG sequence classes, different in methylation status, were observed in rice genome: Methylation status at the class 1 CG sites was conserved among genetically diverse rice cultivars, whereas cultivar-specific differential methylation was frequently detected among the cultivars at the class 2 CG sites. Five class 2 CG sites were localized on different chromosomes and were not clustered together in the genome; the differential methylation was stably inherited in a Mendelian fashion over 6 generations, although alterations in the methylation status at the class 2 CG sites were observed with a low frequency (Ashikawa 2001).

Usually the individual plant genes and corresponding promoters are methylated quite unevenly. In *Silene latifolia* a male reproductive organ-specific gene *(MROS1)* expressed in the late phases of pollen development is very intensively methylated at CG sites (99%) in the upstream region, whereas only a low level of CG methylation (7%) was observed in the transcribed sequence; the asymmetric sequence methylation (2%) in both regions is quite similar (Janousek et al. 2002). The methylation patterns of cytosine residues in the *Arabidopsis thaliana* gene for domain-rearranged methyl-transferase (DRM2) were studied in wild-type and several transgene plant lines containing antisense fragments of the cytosine DNA methyltransferase gene *METI* under the control of copper-inducible promoters (Ashapkin et al. 2002).

It was shown that the promoter region of the *DRM2* gene is mostly unmethylated at the internal cytosine residue in CCGG sites, whereas the 3'-end proximal part of the gene-coding region is highly methylated. Cytosine methylation in CCGG sites in the *DRM2* gene are variable between wild-type and different transgenic plants. The induction of antisense *METI* constructs with copper ions in transgene plants in most cases leads to further alterations in the *DRM2* gene methylation patterns (Ashapkin et al. 2002).

## REPLICATIVE DNA METHYLATION AND DEMETHYLATION

DNA synthesis in L cells and tobacco cells at a relatively high cell concentration (2-4×$10^5$ cells/ml) in a medium is mainly limited to formation of Okazaki fragments (Vanyushin 1984). Thus, it was a unique opportunity to isolate and investigate the character and level of methylation of the Okazaki fragments accumulated. It was shown that these fragments do contain $m^5C$ (Bashkite et al. 1980; Vanyushin 1984), providing evidence that replicative DNA methy-lation, which starts even at the very early stages of replication, does exist in plants and animals. The level of methylation of Okazaki fragments was about twofold lower compared with that of ligated, newly formed and mature DNAs.

The distribution pattern of $m^5C$ among pyrimidine clusters isolated from the Okazaki fragments and ligated DNA was different. Methylation of the Okazaki fragments was relatively insensitive to methylation inhibitor *S*-isobutyladenosine in L cells and to plant growth regulator auxin (2,4-D) in tobacco cells (Bashkite et al. 1980; Vanyushin 1984) whereas methylation of ligated DNA was blocked by these agents. Thus, even early replicative DNA methylation proceeds through at least two phases that may be served by DNA methyltransferases different in site specificity and sensitivity to various modulators. In tobacco cells, another inhibitor of DNA methylation ethionine, unlike 5-azaC, strongly inhibits methylation of cytosine residues in CCG but not CG sequences (Bezdek et al. 1992). The methylation of cytosine residues in CCG and CAG in plant cells is more sensitive to suppression by AdoHcy and is under more stringent AdoHcy/AdoMet control compared with CG methy-lation (Fojtova et al. 1998). Dihydroxypropyladenine (a potential inhibitor of DNA methyltransferase activities by increasing the *S*-adenosylhomocysteine level) induces, in tobacco repeats, a decrease in methylated sequences in the direction $m^5Cm^5CG$'!$Cm^5CG$'!CCG (Kovarik et al. 2000a, b).

The replicative DNA methylation was observed bothin cell suspension cultures and various organs of an intact plant (Vanyushin 1984; Vanyushin and Kirnos 1988). Cereal seedlings are unique and a very useful model for investigation of replicative and post-replicative DNA methylations in plants. Their growth may be easily synchronized and at least five cycles of synchronous replication of nuclear (n)DNA were observed in an initial leaf during the first 7-dayperiod of the seedlingdevelopment (Kirnos at al. 1983a, b). Coleoptile in cereals functions for a relatively short period at the early stage of ontogenesis, and it dies quickly as the seedling grows and develops. Global nDNA synthesis in coleoptile ceased after a few synchronous replication cycles, and this cessation seems to correspond to the beginning of apoptosis in non-dividing cells (Kirnos at al. 1983b; Vanyushin et al. 2004).

Discrete peaks of total DNA synthesis in entire leaf at the early stage of wheat seedling development seem to correspond to cell cycles in the basal meristematic leaf area. It is very useful for a biochemist, as it allows him in terms of DNA to consider an entire organ in an intact developing plant organism as a single cell and to investigate what happens, in particular, with DNA methylation in a cell cycle (Kirnos et al. 1984, 1986, 1988, 1995). Contrary to the initial leaf, in coleoptile the nuclear-DNA content increase stopped on the fourth day of the seedling life.

Thus, the stop of the nDNA (ñ = 1.700 g/cm$^3$) synthesis in coleoptile is strictly arranged temporally in a program of the early stage of seedling development (Kirnos et al. 1983b). This is an obligatory beginning step of apoptosis and organopto-sis. There is no nDNA replication. Only mtDNA (ñ = 1.718 g/cm$^3$) continues to be very intensively synthesized in coleoptile. Therefore, the aging wheat coleoptiles are a good source for mass plant mtDNA. We failed to detect m$^5$C in wheat mtDNA but have detected m$^6$A in it (Vanyushin et al. 1988).

In wheat seedlings (Kirnos et al. 1984b), as in a suspension culture of tobacco cells (Bashkite et al. 1980), the Okazakifragments are methylated. The methylation level (ML) [100 m$^5$C/(C + m$^5$C) = 7.4±0.5] of Okazaki fragments (<5S) in etiolated seedlings was three to four times lower than that in total wheat nDNA.

After ligation of Okazaki fragments, leading to formation of long replication intermediate fragments (RIF) (8S, e”12S), the ML remained at almost the same level as the Okazaki fragments; therefore, recently replicated DNA is significantly undermethylated. In ligated (e”12S) and mature nDNA, up to 40% of all the m$^5$C residues are located in the Pu-m$^5$C-Pu sequences, whereas in the Okazaki fragments this sequence contains only 20% of all the m$^5$C (Kirnos et al. 1984b). This again suggested that there is a DNA methyltransferase associated with the replication fork that is different from the one methylating the long RIF.

DNA duplexes formed during replication exhibit sharply pronounced asymmetry of the $m^5C$ distribution along the complementary—parent and daughter—DNA chains (Kirnos et al. 1984b). This asymmetry remains in the interphase nuclei and it disappears up to the end of cell cycle (Kirnos et al. 1984b). Based on this observation, a model for regulation of DNA replication by methylation in eukaryotes (plants) was first suggested (Kirnos et al. 1984b, 1988;Vanyushin 1988).According to this model,onlythe symmetrically(fully) methylated DNA duplexes are permitted to be replicated. So, in the early S-phase the completely methylated genome compartments ($S_E$ DNA) may be replicated. In contrast, nucleotide sequences that should enter into replication in the late S-phase ($S_L$-DNA) are methylated asymmetrically and their replication in $S_E$ phase is prohibited. With the termination of the $S_E$-DNA replication, the newly formed $S_E$ duplexes are distinctly asymmetric as to the $m^5C$ content in complementary DNA strands; their transcription seems to be permitted but repeated replication in the same cell cycle is prohibited.

As a result of the persistent process of post-replicative methylation (Kirnos et al. 1984a, 1987, 1988), the $S_L$ sequences from the preceding cell cycle gradually become symmetrically methylated; therefore, the transcription of corresponding (late) genes is terminated and they enter into replication. By the onset of a new S-phase, $S_E$- and $S_L$-DNA sequences will be methylated to the same extent as before the preceding cycle of DNA synthesis. $S_E$ and $S_L$ duplexes attain this level dependingon the rate of post-replicative DNA methylation, co-ordinated with the duration of the cell cycle. Thus, the periodic modulation of the asymmetry of methylated sites in nDNA in sequential cell cycles, via replication and replicative or post-replicative methylation, were regarded as a mechanism regulating the periodicity and fidelity of gene replication in the cell cycle (Vanyushin 1988).

A similar mechanism of regulation of DNA replication by methylation was later shown to exist in bacteria (Bae et al. 2003; Fujikawa et al. 2004) where the replication of fully dam-methylated compartments is permitted but replication of hemimethylated ones is blocked. The *Escherichia coli* SeqA protein recognizes the 11 hemimethylated $Gm^6ATC$ sites in the oriC region of the chromosome, and prevents replication over-initiation within one cell cycle. SeqA (SeqA71-181) specifically binds to hemimethylated DNA containing a sequence with a mismatched $m^6A$:G base pair [$Gm^6A$(:G)TC] as efficiently as the normal hemimethylated $Gm^6A$(:T)TC sequence (Fujikawa et al. 2004). As hemimethylated DNA has unusual backbone structure and a remarkably narrow major groove, these dynamic and structural features provide insights into the specific recognition of hemimethylated GATC sites by the SeqA protein (Bae et al. 2003).

Thus, replication is a main mechanism of formation of demethylated or, as it should be said more carefully, undermethylated or hemimethylated DNA. DNA can be unmethylated due to interfering with maintenance methylation or demethylated by the active elimination (excision) of $m^5C$ residues or even by direct removal of methyl group from $m^5C$. DNA undermethylation by interference with the remethylation of newly replicated DNA should be a slow process. Cui and Fedoroff (2002) have developed an assay that permits rapid demethylation of the *Spm* sequence to be controlled by inducing the expression of the TnpA gene for maize *suppressor-mutator* transposon-encoded TnpA protein.

TnpA is a weak transcriptional activator, and deletions that abolish its transcriptional activity also eliminate its demethylation activity. Demethylation is associated with the formation of a transcription initiation complex, while cell cycle and DNA synthesis inhibitors interfere with TnpA-mediated *Spm* demethylation. TnpA has a much lower affinity for fully methylatedthan for hemimethylated or unmethylated DNA fragments derived from *Spm* termini; it was suggested that TnpA binds to the post-replicative, hemimethylated *Spm* sequence and promotes demethylation either by creating an appropriate demethylation substrate or by itself participating in or recruiting a demethylase (Cui and Fedoroff 2002).

Active DNA demethyla-tion in plants was observed during pollen development (Oakeley et al. 1997) and vernalization (Sheldon et al. 1999); progression of tubers through dormancy is accompanied by decreases in methylation at 5-CCGG-3' sequences in potato meristem (Law and Suttle 2003). Strong strand-biased DNA methylation character was observed in heterochromatic *Arabidopsis* centromeres. Unlike the hemimethylation that occurs when methylated DNA is replicated, the patterns are characterized by nearly complete modification of one strand and limited modification of its complement. As methyltransferases capable of biased modification of complementary strands are yet unknown, this DNA methylation pattern can be associated with:

(1) specific binding of de novo methyltransferases that processively modify one strand,

(2) assembly of centromere-binding proteins that limit methyltransferase access to one strand of newly replicated DNA, or

(3) differential access of methyltransferases to the leading or lagging strand during DNA synthesis (Luo and Preuss 2003).

Like in a cell suspension culture, phytohormones mostly inhibit replicative DNA methylation in wheat seedlings (Kirnos et al. 1986). The strongest (up to 50%) inhibition of replicative DNA methylation was observed in SM and SL phases of the cell cycle. A weak, stimulatory effect was exerted by plant growth regulators 6-benzylaminopurine, 2,4-D, gibberellin and kinetin during prolonged (20 h) incubation of cut-off shoots (Kirnos et al. 1986; Vanyushin 1988). Thus, modulation of DNA methylation is to be one of the molecular mechanisms of phytohormone action in plant cell.

Post-replicative DNA methylation (Kirnos et al. 1984a, 1986, 1987, 1988) and demethylation take place also in plants. In *Silene latifolia,* a rapid decrease in the global DNA methylation level occurs in the cotyledons and hypocotyls during seed germination. This DNA demethylation seems to be non-replicative since it occurred before cell division had begun (Zluvova et al. 2001).

## Cytosine DNA Methyltransferases

When it was clearly shown that $m^5C$ in plant DNA may appear in the different sequences such as CG and CNG (Kirnos et al. 1981; Gruenbaum et al. 1981), the idea of the possible multiplicity of DNA methyltransferases in the nucleus of the plant cell appeared (Kirnos et al. 1981). It was already hard to believe that cytosine residues located in these different DNA sequences may be recognized and modified by the same enzyme. Besides, it was found that in plant (Bashkite et al. 1980) and animal (Demidkina et al. 1979) cells the methylation of Okazaki fragments, in contrast to mature DNA methylation, was relatively insensitive to competitive inhibitors of the DNA methylation reaction (SIBA and others) and plant growth regulators (auxin and others). Also, the distribution pattern of $m^5C$ among pyrimidine isopliths from these fragments and mature DNA was very different (Vanyushin 1984).

These facts led to the conclusion that at least two DNA methyltransferases, different in site specificity and sensitivity to various effectors, should be present in a nucleus (Kiryanov et al. 1982). In addition, the data on the different nature and character of DNA methylation in mitochondria and nuclei in plants (Vanyushin et al. 1988) and animals (Vanyushin and Kirnos 1974) indicated that DNA methyltransferases operating in the nucleus and mitochondria are different. Then it was shown that plant DNA methyltransferases may differ from respective animal enzymes (Theiss et al. 1987; Vlasova et al. 1996), and, in addition to CG methylating activity, the enzymes that preferentially methy-late cytosine in CNG sequences were isolated from pea (Pradhan and Adams 1995) and wheat plants (Vlasova et al. 1995). Now it is clear that the system of cytosine DNA modification in plants is quite complicated and is represented by a family of phylogenetically related but chemically distinct and target-specific DNA methyltransferases (Finnegan and Dennis 1993; Genger at al. 1999; Finnegan and Kovac 2000; Wada et al. 2003).

There are at least three types of DNA methyltransferases in plants: METI, chromomethylase (CMT) and DRM.

The first plant gene *METI* encoding a cytosine methyltransferase was isolated from *Arabidopsis thaliana* (Finnegan and Dennis 1993). Reduction of CG methylation in *met1-1* mutants was associated with developmental abnormalities (Kankel et al. 2003). *METI* genes have been identified also in carrot, pea, tomato and maize (Bernacchia et al. 1998; Pradhan et al. 1998). In fact, *METI* is a member of a multigene family, with up to five members (Finnegan and Dennis 1993; Genger et al. 1999). Four genes arose from an ancestral gene, and the gene structure, including the position of the 11 introns, is conserved between thefamily members (Finnegan and Kovac 2000). The unlinked genes,

*METIIa* and *METIIb*, are products of the most recent gene duplication. *METI* is the predominant methyltransferase in *Arabidopsis* (Genger et al. 1999) and other plants; it preferentiallymethylates cytosineresidues in CG with a highest activity in meristematic cells (Ronemus et al. 1996). *METIIa* and *METIIb* are transcribed in all tissues, but the level of transcript is verymuch lower than for *METI* (Genger et al. 1999). The function of the proteins encoded by *METIIa*, *METIIb*, and *METIII* is unknown; antisense constructs against *METIIa* have no effect on global methylation or plant development. A *METI* antisense did not affect expression of *METIIa/b*, and yet these enzymes were unable to substitute (completely) for METI activity in *METI* antisense plants (Genger et al. 1999).

METI enzymes lack the cysteine-rich zinc-binding region found in the aminoterminal domain of mammalian enzymes (Bestor 1992) and have an acidic region, consisting of at least 50% glutamic acid and aspartic acid residues not found in mammalian Dnmt1-like enzymes (Finnegan and Kovac 2000). It was suggested that similarly to animals the aminoterminal domain in METI is important for discrimination between hemimethylated and un-methylated DNA, giving the enzyme a strong preference for a hemimethylated template to effectively accomplish maintenance methylation (Finnegan and Kovac 2000). The expression of *MET1* is associated with DNA replication: In maize the transcripts of *MET1* exclusively accumulate in actively proliferating cells of the meristems in mesocotyls and root apices (Steward at al. 2000). *METI* antisense decreased methylation of cytosine residues in CG and CCG but not in CAG or CTG sequences (Finnegan et al. 1996).

A cDNA encoding a DNA methyltransferase, with a predicted polypeptide of 1,556 amino acid residues containing all motifs conserved in this enzyme family, was isolated from tobacco plants, and the corresponding gene was designated as *NtMET1*. Similarly to MET1 the NtMET1 transcripts accumulate in dividing tobacco cells and are localized exclusively in actively proliferating tissues around axillary apical meristem. Methylation levels of genomic DNA from transgenic plants with NtMET1 antisense significantly decreased in comparison with wild-type levels, and distinct phenotypic changes including small leaves, short internodes and abnormal flower morphology were noted (Nakano et al. 2000). METI and chromatin remodelling protein DDM1 are required for maintenance of global cytosine methylation of genome in plants (Bartee and Bender 2001).

A second class of methyltransferases—chromomethylases (CMT family)— found in *Arabidopsis* (Henikoff and Comai 1998; Genger et al. 1999) and other plants is characterized by insertion of a chromodomain between conserved motifs II and IV of the methyltransferase domain. Chromomethylases seem to be involved in modifying DNA in heterochromatin, and they are responsible for maintenance of cytosine methylation at CNG sites, particularly in retrotransposons (Lindroth et al. 2001; Tompa et al. 2002). In *Arabidopsis*, CMT3 takes part in methylation of the *SUPERMAN* gene and is responsible for maintaining epigenetic gene silencing; *cmt3* mutants display a wild-type morphology but exhibit decreased CNG methylation of the *SUPERMAN* gene and of other sequences throughout the genome; they also show reactivated expression of endogenous retrotransposon sequences (Lindroth et al. 2001).

Conserved motifs in CMT are relatively (up to 70%) homologous to that of METI; but the length of the aminoterminal domain in CMT proteins is variable, and this domain has no similarity to that of the METI family (Genger et al. 1999). A cytosine DNA methyltransferase containing a chromodomain, *Zea* methyltransferase 2 (ZMET2), was recently cloned from maize. The sequence of ZMET2 is similar to that of the *Arabidopsis* chromomethylases CMT1 and CMT3, and the enzyme is required for in vivo methylation of CNG sequences (Papa et al. 2001). *Arabidopsis cmt3* chromomethylase mutations block non-CG methylation and silencing of an endogenous reporter gene and reduce CNG methylation at repetitive centromeric sequences (Bartee et al. 2001). CMT methyltransferases seem to be unique to plants because no methyltransferases of this class have been identified in species from other kingdoms (Genger et al. 1999).

The third class of methyltransferase genes—composed of *DRM1* and *DRM2*—has catalytic domains with a sequence homologous to those of mammalian Dnmt3 methyltransferases. In a plant (*Arabidopsis*) genome, the sequences homologous to de novo methyltransferases Masc1 from *As-cobolus* and Dnmt3 from mouse are observed (Finnegan and Kovac 2000). The *DRM* loci in plants are required for asymmetric

DNA methylation. At some loci, *drm1drm2* double mutants eliminated all asymmetric methyla-tion, but at the *SUPERMAN* locus this methylation was completely eliminated only in the *drm1drm2cmt3* triple mutant plants.

DRM and CMT3 methylate the same asymmetrical sites that follow cytosine residue (Cao and Jacobsen 2002; Cao et al. 2003). It is interesting that neither *drm1drm2* double mutants nor the *cmt3* single mutants show morphological defects, the pleiotropic defects in plant development (development and growth retardations, partial sterility) were observed only in *drm1drm2cmt3* triple mutants, probably due to distortions in RNA-directed DNA methylation (Cao et al. 2003).

In animal cells, a novel gene, *Dnmt3L*, encodes a protein that acts as a regulator of DNA methylation rather than as a DNA methylation enzyme; the protein functions as a transcriptional repressor through its ability (like Dnmt3a and Dnmt3b) to associate with histone deacetylase activity (Deplus et al. 2002). It cannot be ruled out that a similar situation with some Dnmt3 genes may take place in plant cells also. In tobacco cells the DRM NtDRM1 was described; the enzyme de novo methylates cytosines in non-CG sequences (Wada et al. 2003). *NtDRM1* is constitutively expressed through the cell cycle and in all tobacco plant tissues.

As a constitutive part of multiple protein complexes, the enzyme may take part in modulation of chromatin structure and thereby methylate particular DNA regions (Wada et al. 2003). DRM enzymes from *Arabidopsis*, maize and tobacco contain the conservative ubiquitin association (UBA) domains (Cao et al. 2000; Wada et al. 2003), which suggests a link between DNA methylation and ubiqui-tin/proteasome pathways.

It is assumed that plant DRMs are controlled in a cell cycle by ubiquitin-mediated protein degradation or (and) the ubiq-uitinization may alter the cellular localization of the DRM proteins due to respective external signals, the cell cycle or transposon or retroviral activity. UBA domains are found neither in other classes of plant DNA methyltrans-ferases nor in mammalian Dnmt3 proteins; therefore, ubiquitin-associated pathway may be restricted to Dnmt3-like methylases in plants (Cao et al. 2000).

## METHYL-DNA-BINDING PROTEINS AND MUTUAL CONTROLS BETWEEN DNA METHYLATION AND HISTONE MODIFICATIONS

It has been well known that DNA methylation influences essentially the interaction of DNA in chromatin with various proteins, including different regulatory factors, histones and others. It may diminish or even prevent specific protein binding to target DNA (Staiger et al. 1989; Inamdar et al. 1991; Ehrlich et al. 1992; Ashapkin et al. 1993; Fisscher et al. 1996; Galweiler et al. 2000; Sturaro and Viotti 2001) or vice versa, an obligatory element for such a binding. In animals, DNA methylation can leadto the recruitment of specific m$^5$C-binding proteins taking part in formation of unique gene silencing complexes (Bird and Wo lffe 1999; Hendrich and Bird 2000; Ballestar and Wo lffe 2001; Jaenisch and Bird 2003; Kimura and Shiota 2003; Kriaucionis and Bird 2004).

Genes for them$^5$CG-binding-domain proteins are foundin plantsalso;they are transcriptionally active and crucial for normal plant development (Berg et al.2003).The *Arabidopsis*genomecontains 12 putativegenesfor suchproteins. These putative proteins were identified and classified into seven subclasses (Zemach and Grafi 2003). AtMBD7 (subclass VI), a unique protein containing a double MBD motif, as well as AtMBD5 and AtMBD6 (subclass IV), specifically bind the symmetrically methylated CG sites (Scebba et al. 2003; Zemach and Grafi 2003); the MBD motif derived from AtMBD6, but not from AtMBD2, was sufficient for binding methylated CG dinucleotides. AtMBD6 precipitated histone deacetylase activity from the leaf nuclear extract.

The examined AtMBD proteins neither bound methylated CNG sequences nor did they display DNA demethylase activity. It is suggested that AtMBD5, AtMBD6 and AtMBD7 arelikelyto functionin *Arabidopsis* plants as mediators of theCG methylation, linking DNA methylation-induced gene silencing with histone deacetylation (Zemach and Grafi 2003). On the other hand, it was mentioned that MBD5 and MBD6, despite their high homology, can be differentiated by their ability to recognize methylated asymmetrical sites (Scebba et al. 2003).

Ten membersofthe*Arabidopsis* gene familyencodingmethyl-CG-bindingdo-main proteins are transcriptionally active, differentially expressed in diverse tissues and at least one, AtMBD11, is crucial for normal development (Berg et al. 2003). This protein showed a strong affinity for DNA independently from the level of methylation (Scebba et al. 2003). Transformed *Arabidopsis* plants with a construct aimed at RNA interference with expression of the AtMBD11 gene, normally active in most tissues, displayed the phenotypic effects such as aerial rosettes, serrated leaves, abnormal position of flowers, fertility problems and late flowering. *Arabidopsis* lines with reduced expression of genes involved in chromatin remodelling and transgene silencing show similar phe-notypes (Berg et al. 2003). These data along with others suggest an important role for AtMBD proteins in plant development.

The methyl-DNA-binding proteins were found in pea (Zhang et al. 1989; Ehrlich 1993), maize (Rossi et al. 1997; Sturaro and Viotti 2001) and carrot (Pitto et al. 2000) cells. The Opaque-2 (O2) protein from the maize endosperm cell extracts binds in vitro to the cytosine-methylated target sequence of the maize O2 promoter with different affinities depending on the methylation status of DNA (CG-methylated, hemimethylated, partially methylated and fully methylated target DNA). Thus, it was hypothesized that DNA methylation modulates, in vivo, the response of the promoter to the cognate transcription factors (Rossi et al. 1997). The dcMBP1 protein from carrot protoplasts binds to symmetrically methylated sequences with high affinity and displays binding properties similar to mammalian MeCP2; protein dcMBP2 has unique binding properties, it binds specifically to $m^5C$ in unconventional CNN and symmetrical CNG sequences and seems to be specific for plants (Pitto et al. 2000).

There is no doubt that a peculiar cross-talk between DNA methylation and histone modifications does exist in eukaryotes. In *Neurospora* the methyla-tion of lysine 9 in histone H3 is critical for cytosine DNA methylation, normal growth and fertility of fungus (Tamaru and Selker 2001). Histones there may be a type of the signal transducers for DNA methylation. On the other hand, in *Arabidopsis* the maintenance CG methylation precedes and directs the hi-stone H3 lysine 9 methylation in heterochromatin (Soppe et al. 2002). It is suggested that DDM1, MET1, H3K9-specific histone methylase and histone deacetylase (H4K16) play an essential role in the formation of heterochromatin directly after replication, and the CG methylation is performed when newly formed nucleosomes are still accessible due to acetylated H4K16.

H3K9 methylation directed by methylated DNA seems to complete heterochromatin assembly (Soppe et al. 2002). Complete removal of CG methylation in an *Ara-bidopsis* mutant null for maintenance methyltransferase (homozygous for *met1* mutant) results in a clear loss of histone H3 methylation at lysine 9 in heterochromatin and heterochromatic loci that remains transcriptionally silent; the loss of both CG methylation and H3K9 methylation at condensed heterochromatic centromers had no effect on their structure (Tariq et al. 2003).

This provides additional evidence that methylation of H3K9 is directed by CG DNA methylation, and the process seems to be transcriptionally independent. In a mutant used with completely erased CG methylation, the methylation at the CNG and CNN sites was reduced only to 57.6% and 73%, respectively (Saze et al. 2003). In *kyp* mutants defective in histone H3 ly-sine 9 methyltransferase, the DNA methylation is affected only at CNG and CNN sites, which suggests that non-CG methylation is controlled by histone methylation (Jackson et al. 2002). Loss-of-function kryptonite alleles resemble mutants in the DNA methyltransferase gene *CMT3;* CMT3 interacts with an *Arabidopsis* homologue of HP1, which in turn interacts with methylated histones (Jackson et al. 2002).

The product of the *ddml* gene is one of the ATP-dependent chromatin remodelling factors that is required to maintain histone H3 methylation patterns and control the DNA methylation level. The gene is responsible for transposon and transgene silencing. Thus, transposon methylation in plants maybe guided by histone H3 methylation (Gendrel et al. 2002). As the H3mK9-dependent DNA methylation is carried out by chromomethylase CMT3 that binds histone methylase via an HP-1-like protein, the loss of DNA methylation in *ddml* maybe due to a reduced association of hetero chromatin with H3mK9 (Gendrel et al. 2002).

Histone and DNA methylations are under thecontrol of ARGONAUTE proteins involved in post-transcriptional RNA-mediated gene-silencing systems and in transcriptional gene silencing in various eukaryotes. In the *Arabidopsis ago4-1* mutant, the silent *SUPERMAN* gene was reactivated and the CNG and asymmetric DNA methylations, as well as histone H3 lysine 9 methylation, were decreased. In addition, the accumulation of 25-nucleotide siRNAs that correspond to the retroelement AtSN1 was observed. Thus, *ago4* and long siRNAs direct chromatin modifications, including histone methylation and non-CG DNA methylation (Zilberman et al. 2003). Histone and DNA methylations in plant cells are well co-ordinated and seem to be interdependent.

It was shown that rRNA gene dosage control and nucleolar dominance utilize a common mechanism. Central to the mechanism is an epigenetic switch in which concerted changes in promoter cytosine methylation and specific histone modifications dictate the on and off states of the rRNA genes (Lawrence et al. 2004). A key component of the off switch is HDT1, a plant-specific histone deacetylase that localizes to the nucleolus and is required for H3 lysine 9 deacetylation and subsequent H3 lysine 9 methylation. It is assumed that cytosine methylation and histone deacetylation seem to be each upstream of one another in a self-reinforcing repression cycle (Lawrence et al. 2004).

Thus, like in animal cells (Nan et al. 1998; Jones et al. 1998; Deplus at al. 2002), the close connection between DNA methylation and histone deacetylation does exist in plants (Aufsatz et al. 2002b). Transgenic plants treated with propionic or butyric acid (inhibitors of histone deacetylases) display increased level of DNA methylation and epigenetic variegation (ten Lohus et al. 1995a). Growth of *Brassica* seedlings in the presence of inhibitor of DNA methylation 5-aza-2'-deoxycytidine or histone deacetylase inhibitors (sodium butyrate and trichostatin A) caused the normally silent underdomi-nant *B. oleracea* rRNA genes to become expressed at high levels. It is assumed that there is a nucleolar dominance mechanism combining DNA methylation and histone modifications to regulate rRNA gene activity (Chen and Pikaard 1997).

Expression of the antisense histone deacetylase *AtHD1* responsible for accumulation of acetylated histones is associated with various developmental abnormalities, including early senescence, ectopic expression of silenced genes, suppression of apical dominance, homeotic changes, heterochronic shift toward juvenility, flower defects and male and female sterility; but it is not accompanied by visible changes in genomic DNA methylation (repetitive DNA sequences, rDNA, a specific locus *SUP*) in the transgenic plants. This suggests that AtHD1 is a global regulator that controls gene expression during development (Tian and Chen 2001).

On the other hand, the *AtHDA6* gene for presumed histone deacetylase is required to maintain the DNA methylation pattern induced by double-stranded (ds)RNA (Aufsatz et al. 2002a, b). Mutations in *AtHDA6* result in loss of transcriptional silencing from several repetitive transgenic and endogenous templates; total levels of histone H4 acetylation are only slightly affected, whereas significant hyperacetylation is restricted to the nucleolus organizer regions that contain the rDNA repeats. This switch coincides with an increase of histone 3 methylation at Lys residue 4, a modified DNA methy-lation pattern and a concomitant decondensation of chromatin. Therefore, *AtHDA6* might play a role in regulating activity of rRNA genes, and this control might be functionally linked to silencing of other repetitive templates and to its previously assigned role in RNA-directed DNA methylation (Probst et al. 2004). Thus, in fact, "methylation meets acetylation" (Bestor 1998).

## RNA-Directed DNA Methylation

In plants, RNA-directed DNA methylation (RdDM) involves de novo methyla-tion of almost all cytosine residues in a region of siRNA–DNA sequence identity. Therefore, RdDM is mainly associated with CNG and non-symmetrical methylations (rare in animals) in protein coding and promoter regions of silenced genes (Wassenegger et al. 1994; Jones et al. 1999; Mette et al. 2000; Wassenegger 2000; Chan et al. 2004). RdDM of cytosine residues specifically occurs along the DNA regions that are complementary to the directing RNA, pointing to the formation of a RNA–DNA duplex, and direct RNA–DNA interaction can act as a strong and highly specific signal for de novo DNA methylation. Dense methylation patterns and the methylation of cytosine residues at symmetric and asymmetric sites are detectable on both DNA strands within these DNA regions. Methylation progressively decreases in the sequences adjacent to the putative RNA–DNA duplex (Mette et al. 2000; Wassenegger 2000). The gene-specific precursor dsRNA, rather than small RNA (smRNA), serves as the gene methylation signal (Mallory et al. 2001; Melquist and Bender 2003).

A promoter dsRNA-mediated transcriptional gene-silencing system associated with induced DNA methylation has been clearly established in tobacco, pea and *Arabidopsis*. The nopaline synthase promoter target gene (NosPro-*NTPII*) is active when the NosPro region is unmethylated; but in the presence of the silencing locus, the NosPro region is specifically methylated in symmetrical (CG and CNG) and non-symmetrical (CNN) cytosines. NosPro dsRNA (transcribed from a NosPro-inverted repeats at the silencing locus and processed to short RNAs consisting of 21–24 nucleotides) triggers de novo methylation in any sequence context within the region of the RNA–DNA sequence identity. It silences the target NosPro in *trans* and contributes to methylation in *cis* of the NosPro copies in the inverted repeats at the silencing locus (Aufsatz et al. 2002a). Removing of NosPro dsRNA results in a loss of non-symmetrical cytosine methylation. MET1 and DDM1 are essential (probably as the chromatin restructuring activities) to RNA-directed DNA methylation as, even in the presence of NosPro dsRNA, the significant loss of NosPro methylation in *met1* and *ddm1* mutants was observed (Aufsatz et al. 2002a).

Partial loss of the coding sequence methylation induced by *ddm1* or *met1* mutations can case a partial loss of RNA silencing (Morel et al. 2000). When RNA silencing is blocked by mutations in the SGS2/SDE1 gene, the CNG methylation is abolished and only a low level of CG methylation was observed (Mourrain et al. 2000). The loss of methylation at both *A. thaliana* and *A. arenosa* centromere repeats, due to expression of dsRNA corresponding to the*A. thaliana* (DDM1) gene, was observed. This indicates that a single RNAi-inducing transgene can dominantly repress multiple orthologs (Lawrence and Pikaard 2003).

*DRM* and *CMT3* methyltransferase genes are involved in the initiation and maintenance of RdDM. Neither *drm* nor *cmt3* mutants affected the maintenance of pre-established RNA-directed CG methylation. However, *drm* mutants showed a nearly complete loss of asymmetric methylation and a partial loss of CNG methylation. The remaining asymmetric and CNG methylation was dependent on the activity of CMT3, showing that DRM and CMT3 act redundantly to maintain non-CG methylation. These DNA methyltransferases appear to act downstream of siRNAs, since *drm1drm2cmt3* triple mutants show a lack of non-CG methylation but elevated levels of siRNAs.

DRM activity is required for the initial establishment of RdDM in all sequence contexts including CG, CNG and asymmetric sites (Cao et al. 2003). RdDM was initiated in 35S-GFP (green fluorescent protein) transgenic plants following infection with plant RNA viruses modified to carry portions of either the 35S promoter or the GFP coding region. Targeting of the promoter sequence resulted in both methylation and transcriptional gene silencing that was inherited independently of the RNA trigger. Targeting the coding region also resulted in methylation, but this was not inherited (Jones et al. 2001). Initiation of RdDM was shown to be *MET1*-independent, whereas maintenance of methylation and transcriptional gene silencing in the subsequent generations in the absence of the RNA trigger was *METl*-dependent. Maintenance of methylation associated with systemic post-transcriptional gene silencing was also found to be *MET1*-independent (Jones et al. 2001).

An essential role of a novel putative chromatin-remodelling protein, DRD1, intheRNA-directed DNA methylation has been established recently (Kanno et al. 2004). This protein belongs to a plant-specific subfamily of SWI2/SNF2-like proteins. In *drdl* mutants, RNA-induced non-CG methylation is almost eliminated at the target promoters, resulting in reactivation, whereas methylation of centromeric and rDNA repeats is unaffected. Thus, unlike the SNF2-like proteins DDM1/Lsh1 and ATRX, which regulate methylation of repetitive sequences, DRD1 is not a global regulator of cytosine methylation. DRD1 is the first SNF2-like protein involved in an RNA-guided, epigenetic modification of the genome (Kanno et al. 2004).

RdDM is associated with establishment and maintenance of transgene silencing and virus resistance. Restoration of transgene activity and susceptibility to plum pox potyvirus (PPV) infection of transgenic *Nicotiana benthamiana* plants in sexual progeny correlated with resetting of transgene DNA methylation. RNA signals, generated either by a silenced nuclear gene or by virus replication, both activate a specific cytoplasmic RNA degradation pathway and induce changes in DNA methylation of homologous nuclear genes that switch them from an active to a silenced status (Guo et al. 1999). A sequence-specific RNA-directed de novo methylation of homologous transgenes has been observed following viroid replication in the nucleus of transgenic plants (Wassenegger et al. 1994). In potato spindle tuber viroid (PSTVd)-infected tobacco plants, this process can potentially lead to de novo methylation of all cytosine residues at symmetrical and non-symmetrical sites within chromosomal inserts that consist of multimers of the 359-bp PSTVd cDNA.

A direct RNA-DNA interaction can act as a strong and highly specific signal for de novo DNA methylation. A minimal target size of about 30 bp is necessary for this methylation (Pelissier and Wassenegger 2000). Upon PSTVd infection, expression of transgene (non-infectious fragments of PSTVd cDNA fused to the 3'-end of the GFP-coding region) was suppressed and thepartialde novo methylation of the transgene was observed. PSTVd-specific siRNA was detected but none was found corresponding to the *gfp* gene; methylation was restricted almost entirely to the PSTVd-specific part of the transgene (Vogt et al. 2004). A *gfp* transgene construct lacking viroid-specific elements was not silenced; nor was de novo methylation detected when it was introduced into the genetic background of the PSTVd-infected plant lines containing silenced GFP/PSTVd transgenes.

The absence of *g/p*-specific siRNAs and of significant methylation within the *g/p*-coding region demonstrated that neither silencing nor DNA methylation spread from the initiator region into adjacent 5'-regions (Vogt et al. 2004). On the other hand, some data showed that RNA-directed silencing and DNA methylation can be spread. Virus vectors carrying parts of a GFP transgene targeted RNA silencing in *N. benthamiana* and *Arabidopsis* against the entire GFP RNA, this indicates that there was spreading of RNA targeting from the initiator region into the adjacent 5'- and 3'-regions of the target gene (Vaistij at al. 2002). Spreading was accompanied by methylation of the corresponding GFP DNA; it also was dependent on transcription of the transgene and on the putative RNA-dependent RNA polymerase, SDE1/SGS2. These findings indicate that SDE1/SGS2 produces dsRNA using the target RNA as a template (Vaistij at al. 2002).

When (1) tobacco plants transformed with a chimeric transgene comprising sequences encoding p-glucuronidase (GUS) and (2) the satellite (sat)RNA of cereal yellow dwarf luteovirus were both infected with potato leafroll lu-teovirus (PLRV), which replicated the transgene-derived satRNA to a high level, the satellite sequence of the GUS/Sat transgene became densely methylated.

Within the satellite region, all 86 cytosines in the upper strand and 73 of the 75 cytosines in the lower strand were either partially or fully methylated. In contrast, very low levels of DNA methylation were detected in the satellite sequence of the transgene in uninfected plants and in the flanking nonsatellite sequences in both infected and uninfected plants. All the sequenced GUS/Sat DNA were hypermethylated, the sequence-specific DNA methylation spread into cells in which no satRNA replication occurred, and this was mediated by the spread of unamplified satRNA and/or its associated 22-nt RNA molecules derived from the satRNA (Wang et al. 2001).

In transgenic pea plants, the infection with cytoplasmically replicating RNA pea seed-borne mosaic virus is accompanied by changes induced in transgene methylation associated with the onset of silencing (Jones et al. 1998).

De novo transgene methylation observed at both symmetric and non-symmetric sites on the DNA preceded the onset of resistance, was restricted to sequences homologous to PSbMV viral RNA and only occurred in plants where the outcome was co-suppression (gene silencing). Thus, cytoplasmic viral RNA can affect methylation of homologous nuclear sequences and it maybe the feedback mechanism between the cytoplasm and the nucleus to control the expression of endogenous genes (Jones et al. 1999). In particular, post-transcriptional gene silencing is considered to be responsible for immunity to viral infection in transformed plants that carry homologous viral transgene sequences.

The most probable and comprehensive scenario of RNA-directed DNA methylation in plants (*Arabidopsis*) have been recently suggested by Matzke and et al. (2004):

(1) In the presence of RNA signals, site-specific DMTases cooperate to establish intermediate levels of de novo methylation at CG and non-CG nucleotide groups within a region of RNA–DNA sequence identity;

(2) the RNA-directed pattern of de novo methylation promotes the recruitment of histone-modifying activities;

(3) histone modifications lead to reinforcement of C(N)G methylation, which can also be maintained in the absence of the RNA trigger.

This sequence of events implies that DNA methylation can be both a cause and a consequence of silencing. This dual role might be attributable to the structural resemblance between short RNA–DNA hybrids, which provide a substrate for de novo methylation, and DNA replication forks, where preexisting epigenetic modifications must be preserved. Depending on their sequence composition, individual promoters appear to vary in their sensitivity to different types of cytosine methylation and rely on different DMTases and histone-modifying enzymes to maintain silencing."

Since RNA-directed regulation of DNA methylation in plants is associated with CNG and non-symmetrical DNA methylations, it seems that much more attention should be paid to the search for a similar, regulation type of DNA methylation associated with gene silencing in animals. After all, CNG methy-lation in animal cells is evident (Woodcock et al. 1987; Marinitch al. 2004; Zhang et al. 2004).

## BIOLOGICAL ROLE OF CYTOSINE DNA METHYLATION

Cytosine DNA methylation controls plant growth and development. Similar to animals (Holliday and Pugh 1975; Razin and Riggs 1980; Bird 1992; Razin 1998), specific cytosine DNA methylation in plants controls practically all genetic processes including transcription, replication, DNA repair and cell differentiation, and it is particularly involved in specific gene silencing and transposition. The epigenetic states of various plant genes associated with methylation are stably inherited through generations (McClintock 1967; Brutnell and Dellaporta 1994; Schlappi et al. 1994;

Jacobsen and Meyerowitz 1997; Kakutani et al. 1999; Riddle and Richards 2002).

The inheritable DNA demethylation may be mainly due to the mutations in the respective genes associated with DNA methylation, or it may be induced by known DNA demethylation agents such as 5-azaCyt. For example, the *ddml* mutation in *Arabidopsis* causes a 70% reduction in genomic m$^5$C content and results in stably transmitted developmental abnormalities including defects in leaf and flower structures and flowering time. Remethylation of sequences hypomethy-latedby *ddml* mutation is extremely slow or nonexistent (Kakutani et al. 1996, 1999). *Arabidopsis* plants transformed with an antisense construct of an *Arabidopsis* methyltransferase cDNA (METI) have reduced cytosine methylation in CG dinucleotides. Removal of the antisense construct by segregation in sexual crosses did not fully restore methylation patterns in the progeny, indicating that methylation patterns are subject to meiotic inheritance in *Arabidopsis.*

Plants with decreased methylation displayed a number of phenotypic and developmental abnormalities, including reduced apical dominance, smaller plant size, altered leaf size and shape, decreased fertility and altered flowering time (Finnegan et al. 1996). Maintenance of CpG methylation is essential for epigenetic inheritance during plant gametogenesis: Depletion of *Arabidopsis* MET1 results in immense epigenetic diversification of gametes. This diversity seems to be a consequence of passive post-meiotic demethylation, leading to gametes with fully demethylated and hemidemethylated DNA, followed by remethylation of hemimethylated templates once MET1 is again supplied in a zygote (Saze et al. 2003).

The DNA methylase inhibitors 5-azaCyt and 5-aza-2'-deoxycytidine inhibited adventitious shoot induction in *Petunia* leaf cultures. Cytosine methylation at CCGG and CGCG sites within a MADS-box gene and a CDC48 ho-mologue, among others, shows strong positive correlation with adventitious shoot bud induction (Prakash et al. 2003). Application of the hypomethyla-tion drugs 5-azaCyt or dihydroxypropyladenine to transgenic tobacco lines resulted in about 30% reduced methylation of cytosines located in a non-symmetrical sequences in the 3'-untranslated region of the neomycin phos-photransferase II (nptII) reporter gene, this hypomethylation was accompanied by up to a 12-fold increase in NPTII protein level (Kovarik et al. 2000b).

5-AzacCyt sharply accelerated apoptotic DNA fragmentation in the coleop-tiles of wheat seedlings exposed to this compound, which can be caused by DNA demethylation and, correspondingly, by derepression and induction of various apoptogenic factors, including, for example, caspases, endonucleases and regulatory proteins (Vanyushin et al. 2002). The treatment of plants with 5-azaCyt is responsible for dwarfism in rice (Sano 2002) and an increased storage protein content in wheat seeds (Vanyushin et al. 1990); both are inherited in few generations. In the transgenic rice seedlings the *bar* gene expression induced by 5-azaCyt treatment disappears in about 20-50 days (Kumpatla and Hall 1998). This means that plants have a tendency and ability to re-establish an initial genome methylation pattern that was distorted by the drug. Treatment with 5-aza-2'-deoxycytidine resulted in the development of altered morphologies in the synthetic allotetraploids of *Arabidopsis* and *Cardaminopsis arenosa* (Madlung et al. 2002).

Expression of a cytosine methyltransferase *MET1* from *Arabidopsis thaliana* as an antisense RNA in transgenic plants resulted in a 34% to 71% reduction in total genomic cytosine methylation in both repetitive DNA and single-copy gene sequences. It was accompanied by altered heterochrony, changes in meristem identity and organ number, female sterility, and a prolonged period of both vegetative and reproductive phases of . development. Thus, DNA methylation is involved in establishing or maintaining epigenetic developmental states in the meristem (Ronemus et al. 1996).

Some developmental abnormalities present in an antisense-*M£77* transgenic line resulted from ectopic hypermethylation of the *SUPERMAN* gene (Kishimoto et al. 2001). *SUPERMAN* gene hypermethylation occurred at ahigh frequency in several mutants that cause overall decreases in genomic DNA methylation. Another floral development gene, *AGAMOUS*, also became hypermethylated and silenced in an *Arabidopsis* antisense-AffiTl line (Jacobsen et al. 2000). Ectopic hypermethylation of specific genes in mutant backgrounds that show overall decreases in DNA methylation may be a widespread phenomenon, and it may resemble a phenomenon observed in cancer cells (Jacobsen et al. 2000). The DNA methylation locus *DDM1* is required for maintenance of gene silencing in *Arabidopsis;* the *ddml* mutation had both an immediate and a progressive effect on PAI2 tryptophan biosynthetic gene *(MePAI2)* gene silencing (Jeddeloh et al. 1998).

DNA methylation controls flowering in plants that are needed in vernalization (exposure to cold) to initiate flowering. Vernalization accompanied by DNA demethylation may be substituted for 5-azaCyt treatment or *MET1* inactivation (antisense) that promotes flowering in vernalization-responsive *Arabidopsis* plants (Burn et al. 1993; Finnegan et al. 1998). DNA methylation regulates transcription of *FLC*, a repressor of flowering (Finnegan et al. 1998).

*FLC* is a key gene in the vernalization response. Plants with high *FLC* expression respond to vernalization by downregulating *FLC* and thereby flowering at an earlier time.

The downregulation of *FLC* by low temperatures is maintained throughout vegetative development but is reset at each generation. A small gene cluster, including *FLC* and its two flanking genes, is co-ordinately regulated in response to vernalization (Finnegan et al. 2004). It is remarkable that foreign genes inserted into the cluster also acquire the low-temperature response. At other chromosomal locations, *FLC* maintains its response to vernalization and imposes a parallel response on a flanking gene; thus, *FLC* contains sequences that confer changes in gene expression extending beyond *FLC* itself, perhaps through chromatin modification (Finnegan et al. 2004).

Cold stress induces DNA demethylation in various plants. In particular, it may be associated with cold-dependent expression of specific proteins. When maize seedlings were exposed to cold stress, a genome-wide demethylation occurred in root tissues (Steward et al. 2002). One particular 1.8-kb fragment (*ZmMI1*) containing a part of the coding region of a putative protein and part of a retrotransposon-like sequence was demethylated and transcribed only under cold stress. Interestingly, coldstressinduced severeDNA demethylation in the nucleosome core but not in the linkers. Methylation and demethylation were periodic in nucleosomes (Steward et al. 2002).

It is known that the transposition frequency ofTam3 in*Antirrhinum majus*, unlike that of most other cut-and-paste-type transposons, is tightlycontrolled by temperature. Tam3 transposes rarely at 25°C, but much more frequently at 15°C. The temperature shift induced a remarkable change of the methy-lation state unique to Tam3 sequences in the genome: Higher temperature resulted in hypermethylation, whereas lower temperature resulted in reduced methylation. The methylation state was reversible within a single generation in response to a temperature shift (Hashida et al. 2003). Differences in the methylation pattern were observed in the DNA of spring and winter wheat (*Triticum aestivum*), as well as in unvernalized and vernalized wheat plants. Winter wheat was more highly methylated than spring wheat; changes in the methylation pattern were observed at the end and after vernalization. Thus, there is not only a vernalization-induced demethylation related to flower induction, but there is also a more general and non-specific demethylation of sequences unrelated to flowering (Sherman and Ta lbert 2002).

DNA methylation in plants is involved in parental imprinting and regulation ofthedevelopmentalprogramme (Finnegan etal. 2000). In sexualspecies, endosperm typically requires a ratio of two maternal genomes to one paternal genome for normal development, but this ratio is often altered in apomicts, suggesting that the imprinting system is altered as well. DNA methylation is one mechanism by which the imprinting system could be altered to allow endosperm development in apomicts (Spielman et al. 2003).

Analysis of inbred lines and their reciprocal crosses in maize identified a large number of conserved, differentially methylated DNA regions (DMRs) that were specific to the endosperm. DMRs were hypomethylated upon maternal transmission, whereas upon paternal transmission the methylation levels were similar to those observed in embryo and leaf. Maternal hypomethylation was extensive and offers a likely explanation for the 13% reduction in $m^5C$ content of the endosperm compared with leaf tissue (Lauria et al. 2004). In the maize endosperm, genes for a-zeins and a-tubulins methylated in sporo-phytic diploid tissues become undermethylated in the triploid endosperm, and the demethylation correlating with gene expression is often restricted to the two chromosomes of maternal origin (Lund et al. 1995a, b).

In *Arabidopsis* the paternally inherited *MEA* alleles are transcriptionally silent in both young embryo and endosperm. *MEA* gene imprinted in the *Arabidopsis* endosperm encodes a SET-domain protein of the Polycomb group that regulates cell proliferation by exerting a gametophytic maternal control during seed development. *ddml* mutations are able to rescue *mea* seeds by functionally reactivating paternally inherited MEA alleles during seed development. Thus, the maintenance of the genomic imprint at the *mea* locus requires zygotic DDM1 activity (Vielle-Calzada et al. 1999). Imprinting of the *MEA* Polycomb gene is controlled in the female gametophyte by antagonism between the two DNA-modifying enzymes, MET1 methyltransferase and DME glycosy-lase (Xiao et al. 2003). DME DNA glycosylase activates maternal *MEA* allele expression in the central cell of the female gametophyte, the progenitor of the endosperm. Maternal mutant *dme* or *mea* alleles result in seed abortion.

Mutations that suppress *dme* seed abortion have been found to reside in the MET1 methyltransferase gene. *MET1* functions upstream of, or at, *MEA* and is required for DNA methylation of three regions in the *MEA* promoter in seeds (Xiao et al. 2003). Parental imprinting in *A. thaliana* involves the activity of the DNA *MET1* gene. Plants transformed with an antisense *MET1* construct have hypomethylated genomes and show alterations in the behaviour of their gametes in crosses with wild-type plants. A hybridization barrier between 2x *A. thaliana* (when used as a seed parent) and 4x *A. arenosa* (when used as apollen parent) can be overcome by increasing maternal ploidy but restoredby hypomethylation.

Thus, hypomethylation restores the hybridization barrier through paternalization of endosperm. Manipulation of DNA methylation can be sufficient to erect hybridization barriers, offering a potential mechanism for speciation and a means of controlling gene flow between species (Bushell et al. 2003).

The *Arabidopsis FWA* gene displays imprinted (maternal origin-specific) expression associated with heritable hypomethylation of repeats around transcription starting sites in endosperm. The *FWA* imprint depends on the maintenance DNA methyltransferase MET1 and is not established by allele-specific de novo methylation but by maternal gametophyte-specific gene activation, which depends on a DNA glycosylase gene, *DEMETER* (Kinoshita et al. 2004).

DNA methylation is essential for genome management in plants: It controls the activity of transposable elements and introduced DNA segments and is responsible for transgene silencing (Kooter et al. 1999; Kumpatla and Hall 1999; Meyer 1999). Methylation of the first untranslated exon and 5'-end of the intron in the maize ubiquitin 1 promoter complex and condensation of the chromatin in regions containing transgenes correlate with transcriptional transgene silencing in barley (Meng et al. 2003).

The homozygous *ddm1* (for decrease in DNA methylation) mutation of *Arabidopsis* results in genomic DNA hypomethylation and the release of silencing in various genes. When the *ddm1* mutation was introduced into an *Arabidopsis* cell line carrying inactivated tobacco retrotransposon Tto1, this element became hypomethylated and transcriptionally and transpositionally active. Therefore, the inactivation of retrotransposons and the silencing of repeated genes have mechanisms in common (Hirochika et al. 2000). A re-markablefeatureofthe*ddm*1mutation is that it induces developmentalabnor-malities by causing heritable changes in other loci. One of the *ddm1*-induced abnormalities is caused by insertion of CAC1, an endogenous CACTA family transposon.

This class of *Arabidopsis* elements transposes and increases in copy number at high frequencies specifically in the *ddm1* hypomethylation background. Thus, the *DDM1* gene not only epigenetically ensures proper gene expression, but also stabilizes transposon behaviour, possibly through chromatin remodelling or DNA methylation (Miura et al. 2001). Robertson's mutator transposons in the *Arabidopsis* genome are heavily methylated and inactive. These elements become demethylated and active in the chromatin-remodelling mutant *ddm1*, which lost the heterochromatic DNA methylation (Singer et al. 2001).

Thus, DNA transposons in plants are regulated by chro-matin remodelling. Since gene silencing and paramutation are also regulated by *DDM1*, the epigenetic silencing is considered to be related to transposon regulation (Singer et al. 2001). Plant S1 SINE retroposons mainly integrate in hypomethylated DNA regionsand aretargeted bymethylases; methylation can then spread from the SINE into flanking genomic sequences, creating distal epigenetic modifications. This methylation spreading is vectorially directed upstream or downstream of the S1 element, suggesting that it could be facilitated when a potentially good methylatable sequence is single stranded during DNA replication, particularly when located on the lagging strand. Replication of a short methylated DNA region could thus lead to the de novo methylation of upstream or downstream adjacent sequences (Arnaud et al. 2000).

DNA methylation influences the mobility of transposons. The influence seems to be associated, particularly, with different affinity for Ac transposase binding to holo-, hemi- and unmethylated transposon ends. In petunia cells, a holomethylated Ds is unable to excise from a nonreplicating vector, and replication restores excision. A Ds element hemimethylated on one DNA strand transposes in the absence of replication, whereas hemimethylation of the complementary strand causes an inhibition of Ds excision. In the active hemimethylated state, the Ds ends have a high binding affinity for the transposase, whereas binding to inactive ends is strongly reduced (Ros and Kunze 2001).

High-frequency transposition of endogenous CACTA transposons in *Arabidopsis* CACTA elements was detected in *cmt3met1* double mutants. Single mutants in either *met1* or *cmt3* were much less effective in mobilization, despite significant induction of CACTA transcript accumulation. Thus, CG and non-CG methylation systems redundantly function for immobilization of transposons (Kato et al. 2003). DNA methylation in the Tam3 end regions in *Antirrhinum* tended to suppress the excision activity, and the degree of methylation was dependent on the chromosomal position (Kitamura et al. 2001).

Paramutation and mutator (*Mu*) transposon inactivation in maize are linked mechanistically (Lisch et al. 2002). A mutation of a gene, modifier of paramutation 1 (*mop1*), which prevents paramutation at three different loci in maize, can reverse methylation of mutator elements. In *mop1* mutant backgrounds, methylation of nonautonomous *Mu* elements can be reversed even in the absence of the regulatory *MuDR* element. *MuDR* methylation is separable from *MuDR* silencing because removal of methylation does not cause immediate reactivation. The *mop1* mutation does not alter the methyla-tion of certain other transposable elements including those just upstream of a paramutable *b1* gene. Thus, the *mop1* gene acts on a subset of epigenetically regulated sequences in the maize genome, and paramutation and *Mu* element methylation require a common factor (Lisch et al. 2002).

Due to known reaction of the oxidative $m^5C$ deamination conjugated with cytosine methylation (Mazin et al. 1985), DNA methylation is an essential mutagenic factor that is responsible for a well-known phenomenon of CG and CNG suppressions that are common for many plant genes (Lund et al. 2003). Thus, DNA methylation is an important factor of plant evolution.

DNA methylation may be essentially modulated by various biological (viral, bacterial fungal, parasitic plant infections) or abiotic factors that may influence plant growth and development. Interestingly, the Chernobyl radiation accident resulted in a global DNA hypermethylation in some plants investigated (Kovalchuk et al. 2003).

Fungal infections most strongly distort methylation in repetitive but not uniquesequences in plant genome(Guseinov andVanyushin 1975). By this method,fungi, viruses andother infective agents may switch over the gene transcription program in the host plant mostly in favour of the respective infective agent. On the other hand, plants are able to modify viral DNA that is not integrated into the plant genome. A few days after inoculation into turnip leaves, the unencapsidated cauliflower mosaic virus DNA was found to be in a methylated state at almost all HpaII/MspI sites (Tang and Leisner 1998).

In fact, proper DNA methylation may stabilize foreign DNA in host plant (Rogers and Rogers 1992). The foreign DNA introduced into barley cells was able to persist through at least two plant generations. Transformation of barley cells was defined by showing initiation of transcription at the proper site on the barley promoter for the chimeric gene in aleurone tissue from both a primary transformant and its progeny, and by tissue-specific expression (aleurone greater than leaf) in the progeny.

This persistence through many multiples of cell division is considered as formally equivalent to transformation, regardless of whether the DNA was chromosomally integrated or carried as an episome, but does not necessarily represent stable integration into the genome, since the foreign DNA was frequently rearranged or lost (Rogers and Rogers 1992).

The foreign DNA was most stable when plasmid DNA used in transformation lacked adenine methylation but had complete methylation of cytosine residues in the CG at Hpa II sites; adenine methylation alone was associated with marked foreign DNA instability. Thus, barley cells have a system that identifies DNA lacking the proper methylation pattern and causes its loss from actively dividing cells (Rogers and Rogers 1992). These intriguing data on foreign DNA methylation in plant cells may resemble a host restriction-modification phenomenon that is common in prokaryotes.

## ADENINE DNA METHYLATION

### $N^6$-Methyladenine in DNA of Eukaryotes

*$N^6$*-Methyladenine ($m^6A$) occurs as aminorbase in DNA ofvarious organisms. It was first detected in *E. coli* DNA 50 years ago (Dunn and Smith 1955). Then it was shown to be obvious in most bacterial DNA (Vanyushin et al. 1968; Barras and Marinus 1989). It has also been found in DNA of algae (Pakhomova et al. 1968; Hattman et al. 1978; Babinger et al. 2001) and their viruses (Que et al. 1997; Nelson et al. 1998), fungi (Buryanov et al. 1970; Rogers et al. 1986), and protozoa (Gutierrez et al. 2000) including *Tetrahymena* (Gorovsky et al. 1973; Kirnos et al. 1980; Pratt and Hattman 1981), *Crithidia* (Zaitseva et al. 1974), *Paramecium* (Cummings et al. 1974), *Oxytricha* (Rae and Spear 1978), *Trypanosoma cruzi* (Rojas and Galanti 1990), and *Stylonychia* (Ammermann et al. 1981). In DNA ofvarious algae, *$N^6$*-dimethyadenine was detected (Pakhomova 1974). About 0.8% of adenine residues are found as $m^6A$ in DNA of the transcriptionally active macronuclei of *Tetrahymena* (Gorovsky et al. 1973; Kirnos et al. 1980). A methylation site is 5'-NAT-3' (Bromberg et al. 1982), and about 3% methylation sites are GATC (Harrison et al. 1986; Karrer and Van Nuland 1998).

**BF. Vanyushin**

The adenine methylated GATC sites are preferentially located in linker DNA, unmethylated sites are generally in DNA of nucleosome cores, and histone H1 is not required for the maintenance of normal methylation patterns (Karrer and Van Nuland 2002). It was suggested that methylated sites may reflect a distribution of nucleosome positions, only some of which provide accessibility to adenine DNA methyltransferase (Karrer and Van Nuland 2002). However, the enzyme methylating adenine residues in *Tetrahy-mena* DNA has not yet been isolated and its amino acid sequence is unknown. DNA of the slime mould *Physarum flavicomum* becomes sensitive to the *DpnI* restriction endonuclease during encystment. This may be due to the appearance of $m^6A$ residues in GATC sequences in this DNA (Zhu and Henney 1990). Early data on the presence of $m^6A$ in mammalian sperm DNA were ambiguous (Unger and Venner 1966), and attempts to detect and isolate this minor base from DNA of many invertebrates and vertebrates were unsuccessful (Vanyushin et al. 1970; Lawley et al. 1972; Fantappie et al. 2001).

Nevertheless, it was judged from the different resistance of animal DNA to restriction endonucleases sensitive to methylation of adenine residues *(TaqI, MboI* and *Sau3AI)* that some genes (*Myo*-D1) (Kay et al. 1994)—steroid-5-a-reductase genes 1 and 2 (Reyes et al. 1997)—of mammals (mouse, rat) might contain $m^6A$ residues. This indirectly suggests that animals may have adenine DNA methyltransferases. It is interesting that addition of $N^6$-methyldeoxyadenosine (MedAdo ) to C6.9 glioma cells triggers a differentiation process and the expression of the oligodendroglial marker 2',3'-cyclic nucleotide 3'-phosphorylase. The differentiation induced by $N^6$-methyldeoxyadenosine was also observed on pheochromocytoma and ter-atocarcinoma cell lines and on dysembryoplastic neuroepithelial tumour cells (Ratel et al. 2001).

The precise mechanism by which modified nu-cleoside induces cell differentiation is still unclear, but it is considered to be related to cell cycle modifications. The incubation of C2C12 myoblasts in the presence of MedAdo induces myogenesis (Charles et al. 2004). It is remarkable that $m^6A$ was detected by a method based on HPLC coupled to electrospray ionization tandem mass spectrometry in the DNA from MedAdo-treated cells (it remains undetectable in DNA from control cells). Furthermore, MedAdo regulates the expression of p21, myogenin, mTOR and MHC. Interestingly, in the pluripotent C2C12 cell line, MedAdo drives the differentiation towards myogenesis only (Charles et al. 2004). These results point to $N^6$-methyldeoxyadenosine as a novel inducer of myogenesis and further extends the differentiation potentialities of this methylated nucleo-side.

$m^6A$ has been found in DNA of higher plants (Vanyushin et al. 1971; Buryanov et al. 1972). It maybe present in plastid (amyloplast) DNA (Ngernprasirtsiri et al. 1988). In wheat seedlings it is present in heavy (ñ = 1.718 g/ $cm^3$) mitochondrial DNA (Vanyushin et al. 1988; Aleksandrushkina et al. 1990; Kirnos et al. 1992a, b). Similar mtDNA containing $m^6A$ were also found in many other higher plants including various archegoniates (mosses, ferns, and others) and angiosperms (monocots, dicots; Kirnos et al. 1992a). The synthesis of this unusual DNA takes place mainly in specific vacuolar vesicles containing mitochondria, and it is a sort of aging index in wheat and other plants (Kirnos et al. 1992b; Bakeeva et al. 1999; Vanyushin et al. 2004).

There is some indirect evidence (based on the comparison of products of DNA hydrolysis with restriction endonucleases *Mbo*I and *Sau*3A) that some adenine residues in zein genes of corn can be methylated (Pintor-Toro 1987). The *DRM2* gene in *Arabidopsis* was found to be methylated at both adenine residues in some GATC sequences and at the internal cytosine residues in CCGG sites (Ashapkin et al. 2002). Thus, two different systems of the genome modification exist in higher plants. It is absolutely unknown how these systems may interact and to what degree they are interdependent. It appears that adenine methylation may influence the cytosine modification and vice versa. Interestingly, the adenine methylation of the *DRM2* gene observed is most prominent in wild-type plants and appears to be diminished by the presence of antisense *METI* transgenes. Since METI does not possess adenine DNA methyltransferase activity, its action on adenine methylation is evidently a secondary effect mediated through adenine DNA methyltransferase or some other factors. Anyway, we have to keep in mind the idea that there may exist a new sophisticated type of interdependent regulation of gene functioning in plants, based on the combinatory hierarchy of certain chemically and biologically different methylations of the genome.

## ADENINE DNA METHYLTRANSFERASES

$m^6A$ is formed in DNA due to the recognition and methylation of respective adenine residues in certain sequences by specific adenine DNA methyl-transferases. Adenine DNA methyltransferases of bacterial origin can also methylate cytosine residues in DNA with the formation of $m^4C$ (Jeltsch 2001). The comparison of protein structures provides evidence for an evolutionary link between widely diverged subfamilies of bacterial DNA $N^6$-adenine methyltransferases and argues against the close homology of $N^6$-adenine and $N^4$-cytosine methyltransferases (Bujnicki 1999–2000).

Enzymatic DNA methylation in prokaryotes and eukaryotes plays an important role in the regulation of many genetic processes including transcription, replication, DNA repair and gene transposition (Razin and Riggs 1980).

It is also an integrative element of host restriction-modification system in bacteria and some lower eukaryotes (Arber 1974).

Adenine DNA methyltransferases of eukaryotes could be inherited from some prokaryotic ancestor. They may be homologous to known prokary-otic DNA-(amino)methyltransferases due to the very conservative nature of DNA methyltransferases in general. ORFs for putative adenine DNA methyl-transferases were found in nuclear but not mitochondrial DNA of protozoa (*Leishmania major*), fungi (*Saccharomyces cerevisiae, Schizosaccharomyces pombe*), higher plants(*A. thaliana*), and animals (*Drosophila melanogaster, Caenorhabditis elegans, Homo sapiens*; Shorning and Vanyushin 2001).

There is nothing currently known about the ORF expression detected or activity of respective eukaryotic proteins encoded in these organisms. The enzymatic activity of these DNA methyltransferases may be very limited as is true, for example, with the transcription of the *Drosophila melanogaster* $C^5$-cytosine-DNA methyltransferase gene [this insect DNA contains an extremely low amount of 5-methylcytosine (Gowher et al. 2000), and the DNA methyl-transferase gene is a component of a ansposon-similar element expressed only in the early stages of embryonic development] (Lyko et al. 2000).

The amino acid sequences of putative eukaryotic DNA-(amino)methyl-transferases (Shorning and Vanyushin 2001) are very homologous to each other, as well as to real DNA-(amino) methyltransferases of eubacteria, hypothetical methyltransferases of archaebacteria and putative HemK-proteins of eukaryotes (Bujnicki and Radlinska 1999). These putative eukaryotic adenine DNA methyltransferases (ORF) share conservative motifs (I, IV) specific for DNA-(amino)methyltransferases and motifs II, III, V, VI and X. Motif I (it takes part in binding of the methionine part of the *S*-adenosylmethionine molecule and is specific for all AdoMet-dependent methyltransferases) was detected in all eukaryotic ORFs found.

The amino acid composition of the catalytic centre in all putative DNA-(amino)methyltransferases is practically the same; it is extremely conservative and does not have any mutations. It seems that if mutations in the catalytic centre of these enzymes occurred, they either would be effectively repaired or the mutants would be lethal. Motifs V, VI and X in eukaryotic ORFs detected are more similar to analogous motifs in DNA-(amino)-methyltransferases from group *g*. In most ORFs detected, the conservative motifs specific for DNA-(amino) methyltransferases occupy less than half of the total amino acid sequence. Six of these ORFs have a relatively large N-terminal part (about 170–200 amino acid residues) located in front of the conservative motifs.

It cannot be ruled out that the gene of the putative DNA-(amino)methyl-transferase is located in a block of genes regulating the replication of mi-tochondrial DNA.

In fully sequenced mitochondrial genomes of eukaryotes (the liverwort *Marchantia polymorpha*, *Arabidopsis thaliana*, sugar beet, the alga *Chrysodidymus synuroideus*) the nucleotide sequences with significant homology to genes of prokaryotic DNA-(amino)methyltransferases were not observed (Shorning and Vanyushin 2001). It is most probable that an enzyme encoded in the nucleus is transported somehow into mitochondria. Putative proteins AAF52125 of *Drosophila melanogaster* and BAB02202 of *Arabidopsis thaliana* might have a signal peptide for mitochondrial transportation on the N-end. Other ORFs for hypothetical DNA-(amino)methyltransferases of eukaryotes do not have distinct signal peptides on the N-end; but, in fact, this does not mean that they do not have them. Signal peptides may be present on the C-end and different from known N-terminal signals may occur (DeLabre et al. 1999).

The first eukaryotic (plant) $N^6$-adenine DNA methyltransferase (*wadm-tase*) isolated was from the vacuolar vesicle fraction of aging wheat coleoptile (Fedoreyeva and Vanyushin 2002). The vesicles appear in plant apoptotic cells, are enriched with $Ca^{2+}$ and contain actively replicating mitochondria (Bakeeva et al. 1999; Vanyushin 2004). In the presence of *S*-adenosyl-l-methionine, the enzyme de novo methylates the first adenine residue in the TGATCA sequence in the single-stranded (ss)DNA or dsDNA substrates, but it prefers single-stranded structures. Wheat adenine DNA methyltrans-ferase is a $Mg^{2+}$- or $Ca^{2+}$-dependent enzyme with a maximum activity at pH 7.5–8.0. About 2–3 mM $CaCl_2$ or $MgCl_2$ in the reaction mixture is needed for the maximal DNA methylation activity.

The enzyme is strongly inhibited by ethylenediaminetetraacetate (EDTA). The optimal concentration of AdoMet in DNA methylation with *wadmtase* is about 10 μM. *Wa dmtase* encoded in the wheat nuclear DNA may be homologous to the *A. thaliana* ORF (GenBank, BAB02202.1), which might be ascribed to putative adenine DNA methyltrans-ferases (Shorning and Vanyushin 2001). The methylated adenine residues foundinGm$^6$AT C s i t e s o fa*DRM2*genein anuclear DNAof*A.thaliana*(Ashap-kin et al. 2002) could be a constituent part of a sequence TGATCA recognized and methylated by wheat adenine DNA methyltransferase. Unfortunately, we do not know whether adenine DNA methyltransferase in *Arabidopsis* cells has the same site specificity as it has in wheat plants.

Since*wadmtase*isfoundinvesicleswithmitochondrialactively-replicating DNA, its maximal activity is associated with mtDNA replication and it prefers to methylate ssDNA, this enzyme seems to operate mainly with replicating mtDNA. Similar to the known dam enzyme controlling plasmid replication in bacteria, *wadmtase* seems to control replication of mtDNA that are repre-sentedmainly by circular molecules in wheat seedlings (Kirnos et al. 1992a, b). As mitochondria could be evolutionarily of bacterial origin, the bacterial control for plasmid replication by adenine DNA methylation seems to be acquired by plant cells, and it is probably used for the control of mitochondria replication.

## PUTATIVE ROLE OF ADENINE DNA METHYLATION IN PLANTS

Unfortunately, the functional role of adenine DNA methylation in plants and other higher eukaryotes is unknown. There are some data available showing that the character of transcription of many plant genes and the morphology and development of transformed plant cells and the plants are drastically changed after introduction into them of genetic constructs with expressed genes of prokaryotic adenine DNA methyltransferases. For example, introduction and expression of the bacterial adenine DNA methyltransferase (*dam*) gene is accompanied by GATC sequence methylation in DNA of transgenic tobacco plants and changes in the leaf and inflorescence morphology. The efficiency of adenine DNA methylation was directly proportional to expression levels of the *dam* construct, and methylation of all GATC sites was observed in a highly expressing line.

Increasing expression levels of the enzyme in different plants correlated with increasingly abnormal phenotypes affecting leaf pigmentation, apical dominance and leaf and floral structure (van Blokland et al. 1998). Moreover, *dam*-methylation of promoter regions in constructs with plant genes for alcohol dehydrogenase, ubiquitin and actin results in an increase in the transcription of these genes in tobacco and wheat tissues (Graham and Larkin 1995). This preliminary methylation of promoters is also important for transcription of *PR1* and *PR2* genes in constructs introduced into tobacco protoplasts by electroporation (Brodzik and Hennig 1998). Adenine methylation of the AG-motif sequence AGATCCAA in the promoter of NtMyb2 (a regulator of the tobacco retrotransposon Tto1) by bacterial dam methylase enhances activity of the AG-motif-binding protein (AGP1) in tobacco cells (Sugimoto et al. 2003). The presence of methylated adenine residues in the sequence GATC scattered in the reporter plasmid introduced into intact barley aleurone layers by a particle bombardment increased transcription from hormone-regulated á-amylase promoters two- to fivefold, regardless of the promoter strength, and proper hormonal regulation of transcription was maintained (Rogers and Rogers 1995). The methylated adenine effect was similar when the amount of reporter construct DNA used was varied over a 20-fold range, beginning with an amount that gave only a small increment of expression.

Similar transcription-enhancing effects for methylated adenine residues in DNA were seen with the CaMV 35S, maize Adh1 and maize ubiquitin promoters (Rogers and Rogers 1995). It was shown that some proteins present in wheatgerm nuclear extracts bound preferentially to adenine-methylated DNA rather than cytosine-methylated DNA. It seems that enhanced transcription of nuclear genes in barley due to the presence of $m^6A$ residues in the vicinity of active promoters may be mediated by $m^6A$ DNA-binding protein (Rogers and Rogers 1995).

Hence, methylation of adenine residues in DNA may control gene expression in plants. This all means that adenine DNA methylation in plants is not an incidental or unexpected event, and it may play a significant physiological role. It was hypothesized that modulation of methylation of adenine residues by incorporation of cytokinins ($N^6$-derivatives of adenine) into DNA may serve as a mechanism of phytohormonal regulation of gene expression and cellular differentiation in plants (Vanyushin 1984). Cytokinins (6 benzylaminopurine) can incorporate into the DNA of plants (Kudryashova and Va n yushin 1986) and *Te t r a hymena pyriformis* (Mazin and Va n yushin 1986). In fact, 6-benzylaminopurine inhibits plastid DNA methylation in sycamore cell culture and induces in these cells the expression of enzymes involved in photosynthesis (Ngernprasirtsiri and Akazawa 1990). It cannot be ruled out that in this particular case, cytokinin may be involved in regulation of adenine DNA methylation in a plastid.

The data showing that adenine DNA methylation may be involved in a control for persistence of foreign DNA in a plant cell is of special interest. Unlike cytosine methylation, the adenine methylation alone is associated with marked foreign DNA instability (Rogers and Rogers 1992). Plant cells seem to have a system discriminating between adenine and cytosine DNA modifications, andthe specific enzymes resemblingto some extent bacterialrestriction endonucleases could be responsible for selective elimination of impropriate adenine methylated DNA. Recently we have isolated from wheat seedlings a few specific AdoMet-, $Ca^{2+}$,

$Mg^{2+}$-dependent endonucleases discriminating between methylated and unmethylated DNAs (Fedoreyeva and Vanyushin 2004; B.F. Vanyushin, unpublished). This may also indicate on the presence of R-M system in higher plants.

## Conclusions

DNA methylation controls plant development and is involved in gene silencing and parental imprinting. It takes part in control for transgenes and foreign DNA. Severe distortions in DNA methylation are accompanied by essential changes in plant growth and morphology. But unlike animals where *dmt1* knockout results in a block of development and is mostly lethal, plants lacking analogous enzyme MET1 survive. It seems that other, less-specific DNA methyltransferases or specific modifications of proteins surrounding the DNA methylation site may compensate for the absence of MET1. Plants have a system of siRNA gene silencing conjugated with a RNA-directed DNA methylation carried out by enzymes capable of performing CNG and unconventional methylations. This system is considered a mechanism for the control of viral infections and even for plant immunity to viral infections, but the exact mechanisms of these events need to be investigated much further.

There is no doubt that DNA methylation is only an integral part of a complex system in an ensemble of unique structures that control gene activity mostly carried out in chromatin, while being closely interdependent on the histone code. The control of DNA methylation in a cell mayexist at least at three levels:

(1) enzyme(s) activity,

(2) $CH_3$-donors and

(3) availability of the substrate DNA to be modified in a fluctuating chromatin structure.

Some plant DNA methyltransferases are unique, they contain the conservative ubiquitin association (UBA) domain and seem to be controlled in a cell cycle by ubiquitin-mediated protein degradation or (and) the ubiquitiniza-tion may alter the cellular localization of these enzymes due to respective external signals, the cell cycle or transposon (or retroviral) activity.

Along with cytosine methylation, the methylation of adenine in plant DNA was observed and specific adenine DNA methyltransferase was described. The same plant gene may be methylated at both the adenine and cytosine residues. The functional role of adenine DNA methylation is still unknown. Anyway, two different systems of the genome modification based on methylation of adenines and cytosines exist in higher plants. It is yet unknown how these systems may interact and to what degree they are interdependent. It appears that adenine methylation may influence cytosine modification and vice versa, and mutual control for these genome modifications may be a part of the epigenetic control of gene activity in plants.

The specific endonucleases discriminating between DNA methylated and unmethylated at adenine and cytosine residues seem to be present in plants. It means that plants may have a restriction-modification system.

Further investigation of chromatin and the interaction of DNA-modifying enzymes with various factors or proteins, including hormone-receptor complexes, is a most important task towards the resolution of the problem of time, place and role of DNA methylations in a plant cell.

# 5

# De Novo Methylation, Long-Term Promoter Silencing, Methylation Patterns in the Human Genome, and Consequences of Foreign DNA Insertion

This chapter presents a personal account of the work on DNA methylation in viral and mammalian systems performed in the author's laboratory in the course of the past 30 years. The text does not attempt to give a complete and meticulous account of the work accomplished in many other laboratories; in that sense it is not a review of the field in a conventional sense.

Since the author is also one of the editors of this series of *Current Topics in Immunology and Microbiology* on DNA methylation, to which contributions by many of our colleagues in this field have been invited, the author's conscience is alleviated that he has not cited many of the relevant and excellent reports by others. The choice of viral model systems in molecular biology is well founded. Over many decades, viruses have proved their invaluable and pioneering role as tools in molecular genetics. When our interest turned to the demonstration of genome-wide patterns of DNA methylation, we focused mainlyon the human genome.

The following topics in DNA methylation will be treated in detail:

(1) The de novo methylation of integrated foreign genomes;

(2) the long-term gene silencing effect of sequence-specific promoter methylation and its reversal;

(3) the properties and specificity of patterns of DNA methylation in the human genome and their possible relations to pathogenesis;

(4) the long-range global effects on cellular DNA methylation and transcriptional profiles as a consequence of foreign DNA insertion into an established genome;

(5) the patterns of DNA methylation can be considered part of a cellular defense mechanism against foreign or repetitive DNA; which role has food-ingested DNA played in the elaboration of this mechanism?

The interest in problems related to DNA methylation has spread—like the mechanism itself—into many neighboring fields. The nature of the transcriptional programs orchestrating embryonal and fetal development, chro-matin structure, genetic imprinting, genetic disease, X chromosome inactivation, and tumor biology are but a few of the areas of research that have incorporated studies on the importance of the hitherto somewhat neglected fifth nucleotide in many genomes. Even the fly researchers now have to cope with the presence of this nucleotide, in however small quantities it exists in the genome of their model organism, at least during embryonal development. The bulk of the experimental work accomplished in the author's laboratory has been shouldered by many very motivated undergraduate and graduate students and by a number of talented postdoctoral researchers. Their contributions are reflected in the list of references in this chapter. We have also had the good luck to receive funding through a number or organizations as acknowledged.

## Introduction

The results of research on the biochemistry and biology of DNA methylation have grown into a sizable body of scientific information. This series within *Current Topics in Microbiology and Immunology* will provide a summary of experimental work and evolving concepts. A single chapter like this one cannot, ofcourse, even attempt to present an adequate overview of this rapidly developing field. This chapter has therefore been restricted to a synopsis of selected work performed in the author's laboratory between 1975 and 2005. For a long time, many colleagues in molecular biology resisted recognizing the fact that the fifth nucleotide in DNA, 5-methyl-deoxycytidine (5-mC), exerts decisive functions in chromatin structure and in genetic control mechanisms. With 5-mC, however, the arguments have finally become too strong to beignored.

Nevertheless,textbooksstillpreachtheexistence of four, insteadof five, nucleotides in DNA. Of course, it is good and essential scientific practice to cast most critical scrutiny on new claims and demand ample and definitive experimental proof. A large number of researchers have now provided this proof, and many of the findings will be summarized in these volumes. My own group started contributing to the honing of problems related to DNA methylation in the mid-1970s, and this article presents a detailed summary of our results that have been adduced since then and stood the test of time. For further information, the reader can consult the references cited herein and previous reviews that have been published as our work proceeded (Doerfler 1981, 1983, 1995, 1996, 2000; Doerfler et al. 1988, 2001).

The discovery of 5-mC (Hotchkiss 1948) in eukaryotic, particularly in mammalian, DNA has provoked a challenging search for its functional significance. This searchis by no means completed, and active investigations on numerous unsolved questions are still continuing. The modification of cytidine (C) to 5-mC, apparently the only one among the nucleotides in mammalian DNA, is introduced post-replicationally by several DNA methyltransferases (DMTases) that are chosen depending on the functional context of their enzymatic activity: DNA can be methylated de novo, still a most enigmatic series of events, or a given pattern of DNA methylation in the genome can be maintained upon replication. In this latter mode of maintenance methylation, the parental DNA strand with the 5-mC residue still in place can serve as the template to direct the DMTases to modify the newly synthesized DNA complement.

Although several DMTases have been well characterized (for a review, see the chapter by T. Chen and E. Li, this volume), it is not clear whether any one of them by itself suffices to facilitate either of the two modes of DNA methylation. In addition to the enzymatic activity proper, the function of these enzymes seems to depend critically on the conformation of the local chromatin segment in which the DNA is to be methylated. Since our understanding of chromatin structure is incomplete, we cannot expect to obtain a comprehensive description of the enzymatic activities of the DMTases. It appears more realistic to propose a complex interplay between DNA-chromatin structure and specific choices of enzymatic functions in which additional regulatory proteins have to participate. In experimental terms, DNA methylation activities cannot be realistically assessed by relying on the measurement of enzymatic function using a naked DNA template, since the actually operational template for DMTases is a DNA-chromatin complex with site-specific, stochastically malleable functions that are targeted to individual loci in the genome. It will be some time before these processes can be elucidated or even mimicked by current technology.

How can we approach a functional analysis of DNA methylation in eukary-otic, particularly in mammalian, systems? One important parameter in understanding this functional DNA modification is to realize that 5-mC residues are not introduced randomly by a fortuitously acting enzymatic mechanism. In contrast, highly specific patterns in the distribution of 5-mC residues exist all over the genome. These patterns appear to be different in each cell type and in each region of the genome. It will require a major effort to determine these patterns of DNA methylation in all parts of the mammalian, specifically in the human, genome.

In recognizing the very significant accomplishment of determining the nucleotide sequence of the human genome, I submit that the task has not been completed without the inclusion of the fifth nucleotide. Of course, it is technically impossible to differentiate between a C- and a 5-mC-residue by the conventional sequencing reaction. The application of the bisulfite protocol of the genomic sequencing reaction (Frommer et al. 1992; Clark et al. 1994) is a demanding project, particularly when it has to be extended to many kilobases of DNA sequence. Nevertheless, this method is, at least for the time being, the only reliable procedure to ascertain levels and patterns of DNA methylation. By applying the bisulfite reaction, one can detect all 5-mC residues in a sequence.

The human epigenome project has just been initiated on an exploratory basis and will have to cope with the fact that patterns of

DNA methylation can be different from cell type to cell type and, of course, in each segment of the genome (Beck and Olek 2003). In my laboratory, we have investigated methylation patterns in several areas of the human genome to obtain a first impression of the types of patterns (see Sect. 6 of this chapter). The structure of the genome inside its chromatin casing and its regulatory functions appear to depend on these patterns of DNA methylation.

The function of the genome will not be understood before the completion of the analysis of these patterns. Hence, the study of more complex biomedical problems will undoubtedly escape a thoroughly informed experimental approach before this analysis has been finished. Currently available data imply that in 90 human genes in the major histocom-patibility complex (MHC) of multiple tissues and individuals, the majority of regions were hypo- or hypermethylated. The patterns were tissue-specific, interindividually variable, and correlated with gene expression (Rakyan et al. 2004).

The following *Gedankenmodell* may aid the conceptual visualization of a more general function of patterns of DNA methylation across the entire genome. The model is based upon the notion that 5-mC residues are modulators of DNA-protein interactions, as proposed earlier (Doerfler 1983), and these modulators could facilitate and enhance or abrogate such interactions.

The direction in which these modulations work- depends on the type of protein and DNA sequences in functionally crucial interactions.

Imagine a bare wall represented here by the plain nucleotide sequence of A, C, G, and T residues onto which elaborate decorations have to be attached. Chromatin proteins then are the decorations that eventually contribute to the chromatin structures and could be specific for different segments of the genome. Now, we insert into the blank wall the 5-mC "pegs" to which proteins bind or are prohibited from binding. With this first and essential set of DNA-protein interactions, a central genome-associated scaffold will be generated that then will be able to inaugurate further protein and/or RNA assemblies until the final, yet enigmatic, chromatin structure has been established.

Local specificities in this structure will, of course, be determined by the site-specific pattern of DNA methylation that thus assumes a functionally crucial role in this assembly process. There are several, but one particular, problem with this model: It is not apparent whether the generation of a given pattern of DNA methylation arises before or after chromatin formation. Possibly, both events are interdependent and develop concomitantly. Upon DNA replication, an established and inheritable pattern of DNA methylation is, of course, maintained by the array of 5-mC residues that are still preserved after DNA replication in the parental strand and that can serve as a template for the insertion of methyl groups in the newly synthesized DNA complement. In this way, patterns of DNA methylation are propagated and inherited. The methylation patterns in turn promote the site-specific chromatin structures.

A further tantalizing aspect arises from the fact that DNA methylation patterns are erased early in embryonic development and are thereupon re-imposed byan unknown mechanism ofde novo DNA methylationthat cannot avail itself of the template pattern on the complementary strand of DNA. Conversely, the fixation of de novo methylation patterns on integrated foreign DNA or in the course of embryonic development might be directed by local chromatin structures that then would have to be "remembered" even in the absence of the fifth nucleotide. It is this crucial interdependence between methylation pattern and chromatin structure that we cannot yet satisfactorily explain. RNA could conceivably serve as a mediator for this functional gap in time and structure. This model is based on the finding that each individual segment of the genome is tightly associated with a given pattern of DNA methylation and, consequently, of chromatin structure. The same or a very similar site-specific pattern can also be conveyed to foreign DNA subsequent to its insertion into a specific segment of the mammalian genome.

In part, this model has been deduced from the observation that the site-specific re-integration of an unmethylated mouse gene, the B lymphocyte tyrosine kinase (BLK) gene, into the mouse genome by homologous recombination leads to the reestablishment of the original and authentic DNA methy-lation pattern in the integrate at its authentic site (Hertz et al. 1999). In contrast, when the BLK gene randomly hits host DNA sequences and re-combines there by a non-homologous mechanism, patterns of DNA methyla-tion are completely different from the authentic pattern in the BLK gene. For a working hypothesis, we assume that each genome segment is characterized by a "methylation memory." Its biochemical correlate is not known but must somehow be related to topical chromatin structure as well as local DMTase type, concentrations, and activities, as well as auxiliary functions.

The most intensely studied function of DNA methylation in eukaryotic genomes is that of promoter activity and long-term gene silencing. Starting in the late 1970s, our laboratory has regularly contributed to the elaboration of this concept (Doerfler 1981, 1983, for reviews). In conjunction (and again in an interdependent mode), DNA (5-mC) and chromatin (histone acetyla-tion and methylation) modifications collaborate in the long-term silencing of promoters, and thus assume an essential function in regulating the activity of specific genome segments.

In recent years, these mechanisms have been recognized to be of importance also for the understanding of more complex biomedical problems, in particular those that are related to genetic imprinting, embryonic and fetal development, genetic disease, and tumor biology. Here, we have another fine example of how basic research on fundamental mechanisms in molecular genetics can eventually help us understand practical problems in biomedical research. Without turning to the study of simpler experimental systems—e.g., viral models—in the elucidation of promoter silencing by DNA methylation and related histone modifications, it would have been impossible to approach more complex problems in mammalian organisms or in plants.

## THE DE NOVO METHYLATION OF INTEGRATED FOREIGN DNA

### Choice of Experimental Systems

There are many excellent examples documenting that the study of viral systems has led to the discovery of fundamental mechanisms in prokaryotic and eukaryotic molecular genetics. Since, in most instances, virus replication has to rely on the utilization of cellular mechanisms, it cannot be surprising that viruses have been efficiently exploited as

Trojan horses inside the cellular milieus. In the 1970s, our laboratory was involved in detailed analyses on the mode of adenoviral DNA integration into the host genome in adenovirus-transformed cells and in adenovirus type 12 (Ad12)-induced hamster tumor cells. In the course of these studies, it became feasible to prove that integrated foreign (adenoviral) DNA was de novo methylated (Sutter et al. 1978). When we subsequently were able to document the first inverse relationship between genetic activity of integrated viral genes and the extent of their methylation (Sutter and Doerfler 1980), it was obvious that this experimental system could be applied to fundamental studies on the regulation of genetic activity and on the biological function of DNA methylation (Doerfler 1983).

A second seminal observation, which emphasized the biological importance of DNA methylation, came from detailed investigations on patterns of DNA methylation in the 5'-upstream regions of two randomly chosen human genes, tumor necrosis factor (TNF)-a and TNF-p. The genomic sequencing method originally developed by Church and Gilbert (1984), though difficult to use at the time, helped considerably in these studies. In the promoter and 5'-upstream regions of the TNF-a and TNF-p genes, we found cell type-specific and interindividually highly conserved patterns in the distribution of 5-mC residues that agreed to the nucleotide site among individuals from different ethnic origins (Kochanek et al. 1990,1991).

Similar, though less precise, evidence came from large, randomly chosen segments of the human genome (Behn-Krappa et al. 1991), many of them repetitive DNA sequences. These results implied that highly specific patterns of DNA methylation existed and most likely had to have a fundamentally important function. It was not easy in those days to convince others that a systematic endeavor to determine patterns of DNA methylation was not just a descriptive exercise but had to be initiated to learn about the wider gamut of possibilities with functional implications. Hopefully, the human epigenome project will help provide more evidence than the study of a single, pioneering, laboratory could possibly have advanced with limited means 15 years ago.

At that time, we also sought the collaboration of clinical researchers in order to extend the basic concepts derived from simpler experimental systems to more complex biomedical problems. In the course of these studies, it became even more obvious that the model developed with the adenovirus system could reliably guide all our efforts. In collaboration with several groups, we determined patterns of DNA methylation in the Prader-Willi/Angelman regions of the human genome (Zeschnigk et al. 1997a, b; Schumacher et al. 1998), in the promoter regions of the RET protooncogene (Munnes et al. 2000), of the FMR1 gene (Schwemmle et al. 1997; Genç et al. 2000) and of several genes of the erythrocyte membrane (Remus et al. 2001,2005).

## THE STATE OF METHYLATION IN DNA VIRAL GENOMES

### Many DNA Virion Genomes Are Unmethylated, Others Are Methylated

Modulation as a bidirectionally active parameter in DNA-protein interactions can be exemplified by the activity of the restriction endonucleases DpnI and DpnII. DpnI cleaves the nucleotide sequences $G^{6m}ATC$ only when the A residue in the recognition sequence is methylated, whereas DpnII is inhibited by a$^{6m}$ A residue in this sequence (for review, see McClelland and Nelson 1988). Hence, a methylated nucleotide can obstruct or facilitate while being required for the activity of a restriction endonuclease, i.e., for the interaction between a given nucleotide sequence and the protein that specifically recognizes this sequence. A similarly instructive example is not available for a 5-mC-containing recognition sequence of a restriction endonuclease.

Among the DNA containing viral genomes, examples of completely unmethylated as well as completely 5'-CG-3'methylated virion DNA molecules exist. The encapsidated virion DNA of the human adenoviruses (Günthert et al. 1976) is unmethylated. In striking contrast, the double-stranded genome of frog virus 3 (FV3), an iridovirus, is completely methylated in all 5'-CG-3' dinucleotides (Willis and Granoff 1980; Schetter et al. 1993). As will be discussed later, the intracellular FV3 virion DNA becomes quickly remethylated after replication. Possibly, due to the specific nucleotide sequence of the FV3 genome, the viral and/or cellular proteins, which have to interact with this viral genome in the course of viral transcription and replication, are not inhibited by FV3 DNA methylation. Some of them may even require a methylated genome for full activity.

We have used several techniques—including total hydrolysis of the virion DNA followed by bidirectional chromatography and electrophoresis (Günthert et al. 1976)—that allow the separation of C from 5-mC residues, as well as genomic sequencing methods (Wienhues and Doerfler 1985; Kämmer and Doerfler 1995), to demonstrate that the virion DNA as well as the free, i.e., not host cell genome integrated, intracellular adenovirus DNA in productively or abortively infected cells (Vardimon et al. 1980) remains unmethylated. In the latter study, restriction endonucleases were used to document the absence of 5-mC residues at least in the HpaII recognition sequences 5'-CCGG-3'.

The intracellular genomes of the episomally persisting Epstein-Barr virus (EBV) have become methylated to a certain extent (Ernberg et al. 1989). Similarly,the genome of another persisting virus, herpesvirus saimiri, in lymphoid tumor cell lines has been shown to be extensively methylated (Desrosiers et al. 1979).

The retroviral progenomes can also become methylated (early references on this topic are e.g., Conklin et al. 1982; Jähner et al. 1982)

## SYREC, an Ad12 Recombinant Genome That Carries Unmethylated Cellular DNA

When Ad12 was serially propagated on human cells in culture, a variant Ad12 genome arose that constituted a naturally generated recombinant between the left terminal 2,081 nucleotides of Ad12 DNA and a large palindromic fragment of cellular DNA. This viral recombinant could be separated from the authentic Ad12 virions due to its lower buoyant density by equilibrium sedimentation in CsCl density gradients (Deuring et al. 1981). The existence of this symmetrical recombinant (SYREC) proved that recombination could occur between viral and cellular DNA in human cells that have been productively infected with human Ad12 (Deuring and Doerfler 1983).

The cellular DNA in this huge palindromic genome of some 34 kb with identical left ends of Ad12 on either terminus of the recombinant genome comprised cellular DNA sequences of both the unique and repetitive types. Interestingly, these cellular DNA sequences were completely unmethylated in the virion recom-binant, but the same cellular DNA sequences were highly methylated in the human cellular genome from which they had been originally derived (Deur-ing et al. 1981). This finding demonstrates that free adenovirion DNA remains devoid of 5-mC in the same human cell nucleus in which adenovirion DNA replicates and in which the methylation of cellular DNA is maintained in specific patterns.

Apparently, the cellular DMTases fail to gain access to the free virion DNA, possibly because adenovirus DNA can avail itself of its own, specific, virion genome-encoded mechanism of DNA replication withthe ade-novirus terminal protein (TP), its viral DNA polymerase (pol), and the DNA binding protein (DBP). Alternatively, it is conceivable that free intranuclear adenovirus DNA becomes protected from de novo methylation by binding to specific proteins. Adenoviral DNA replication is at least partly independent of the cellular replication machinery, except for the requirement for nuclear factors I, II, and III that might not be linked to any of the cellular DMTases. On the other hand, the intracellularly located, episomal DNA of EBV must be tightly associated with the replication system for cellular DNA with which it replicates in synchrony. Thus, the EBV episomes might be in close contact with cellular DMTases and become methylated.

The adenovirus SYREC molecule and its ability to replicate in human cells in the presence of a helper adenovirus with an intact authentic viral genome has been the model for the construction of the gutless adenovirus vectors of the third generation (Kochanek et al. 1996b). These researchers have been able to separate the recombinant virus from its wild-type precursor also by equilibrium sedimentation in CsCl density gradients.

## SUPPRESSION OF THE FREQUENCY OF 5-CG-3 DINUCLEOTIDES IN THE GENOMES OF THE SMALL EUKARYOTIC VIRUSES

The dinucleotide 5'-CG-3' is statistically underrepresented in all but four of the small viruses with a genome size of less than 30 kb (Karlin al. 1994). In the larger viral genomes, the abundance of this dinucleotide follows statistical expectations. The retrotransposons in eukaryotic genomes are also characterized by low values of 5'-CG-3' dinucleotides. There are several possible interpretations for these phenomena:

(1) methylation effects during the proviral states of some of these genomes, which would lead to their silencing,

(2) dinucleotide stacking energies,

(3) mutation mechanisms, or

(4) selection during evolution.

## DE NOVO METHYLATION OF FOREIGN DNA THAT WAS INTEGRATED INTO THE MAMMALIAN GENOME

### Studies on Integrated Ad12 Genomes in Transformed or Tumor Cells

In the course of investigations on the mode of Ad12 DNA integration in Ad12-transformed hamster cells by using restriction endonucleases, the de novo methylation of integrated foreign DNA was discovered. The generated fragments of cellular DNA were separated by electrophoresis on agarose gels and further analyzed by Southern blotting (Southern 1975) and hybridization to $^{32}$P-labeled Ad12 DNA or, more specifically, to the $^{32}$P-labeled terminal fragments of Ad12 DNA. In this way, the terminal viral DNA fragments linked to the immediately abutting cellular DNA segments could be identified.

In an attempt to generate small junction fragments that could be more easily analyzed, frequent-cutting restrictases like HpaII were employed. In these experiments, we discovered that the integrated form of Ad12 DNA was not effectively cleaved by HpaII, whereas virion DNA, previously shown to be unmethylated (Günthert et al. 1976), was readily cut. These data implied that the integrated Ad12 DNA had become de novo methylated upon integration into the established hamster genome (see Sutter et al. 1978; Doerfler 1982; Doerfler et al. 1983, for reviews).

This interpretation could be proved when the isoschizomeric restriction endonuclease pair HpaII and MspI became available. Both enzymes recognize the sequence 5-CCGG-3'. MspI cleaves irrespective of the presence of a 5-mC residue in the 3'-located C-position in the recognition sequence, but HpaII is capable of cleaving only the unmethylated sequence (Waalwijk and Flavell 1978). Along these lines, the integrated Ad12 DNA, like virion Ad12 DNA studied as a control, was completely cleaved by MspI, whereas HpaII could cleave only the virion DNA to completion. Integrated Ad12 DNA was cut incompletely by HpaII and was thus recognized to be 5'-CCGG-3' methylated in distinct patterns. Here, we were ableto document one of the early examples for the notion that foreign DNA inserted into established mammalian genomes became heavily methylated (Sutter et al. 1978; Sutter and Doerfler 1980). This now commonly reproduced finding was later extrapolated to numerous other eukaryotic genomes, including those of plants (for review, see Meyer 1995).

The human papillomaviruses (HPVs) 16 and 18 integrated into the genomes of cells from human cervical carcinomas are also methylated in functionally distinct patterns (Badal et al. 2004).

## SITE OF INITIATION OF DE NOVO METHYLATION: SITE OF FOREIGN DNA INTEGRATION IN THE RECIPIENT GENOME

In later studies, we demonstrated in numerous Ad12-transformed hamster cell lines and particularly in Ad12-induced hamster tumor cells that integrated Ad12 DNA is an excellent substrate for the action of cellular DMTases (Kuhlmann et al. 1982a, b; Orend et al. 1991). The patterns generated in different cell lines and tumors exhibited some similarities but did not appear to be identical. Extent and pattern of methylation of integrated foreign DNA were directed rather by the site of integration in the recipient genome than by the nucleotide sequence of the foreign DNA (Orend et al. 1995a), although the latter could have some influence as well.

In this context, the observation was of interest that a cloned E1 segment of adenovirus DNA genomically fixed by transfection into hamster cells and integrated at several different loci in the host hamster genome became methylated to different extents or could remain hypo- or unmethylated (Orend et al. 1995a). Hence, the site of insertion of foreign DNA had to be a strong determinant in its subsequent de novo methylation. The sites of initiation of de novo methylation were determined by using the cloned HindIII DNA fragments of Ad12 DNA as hybridization probes on HpaII- or MspI-cleaved DNA from different Ad12-induced hamster tumors.

These DNA fragments were separated by electrophoresis on agarose gels and subsequently analyzed by Southern blotting and hybridization to the $^{32}$P-labeled Ad12 DNA fragments. The results of a large number of experiments (Orend et al. 1995a) demonstrated that de novo methylation was initiated inside the integrated Ad12 DNA molecules in two paracentrally located regions of the Ad12 genomes. De novo methylation did not commence at or close to the termini of Ad 12 DNA that were linked to cellular DNA. The termini, in fact, remained hypomethylated, possibly because continued hypomethylation and expression of the gene products from the Ad12 regions E1 (left terminus) and E4 (right terminus) were likely selected for in Ad12-transformed cells and in Ad12-induced tumors.

We have investigated the site of initiation of de novo methylation in integrated Ad12 genomes also at the nucleotide level by applying the bisulfite genomic sequencing reaction. When CsCl-purified Ad12,$10^6$-$10^7$ plaque forming units per animal, is injected intramuscularly into newborn hamsters *(Mesocricetus auratus)*, numerous tumors of different sizes develop in the animals' peritoneal cavities. In these intraperitoneal tumors of different sizes, the paracentrally located regions of integrated Ad12 genomes were analyzed by the bisulfite protocol for the state of methylation.

Methylation levels did not exhibit an unequivocal relation to tumor size. Initiation of DNA methylation, moreover, was not emanating from a specific nucleotide or set of nucleotides in the region previously shown to be a site of initiation of de novo methylation. Initiation was rather regional and appeared to emerge from several sites within this region (Orend et al. 1995a; Hohlweg et al. 2003).

Hence, de novo methylation seems to commence in a region and not at a single specific nucleotide. Of course, this experimental approach does not allow us to derive a definite correlation between the time of foreign DNA integration and that of the initiation of DNA methylation.

In a related type of experiment, a plasmid construct, which contained the E2A late promoter of adenovirus type 2 (Ad2) and the prokaryotic gene for the chloramphenicol acetyltransferase (CAT) as a reporter, was transfected into hamster cells. The HpaII-premethylated or unmethylated pAd2E2AL-CAT gene construct was genomically fixed in hamster cells by co-transfection with the unmethylated pSV2-neo plasmid. In this plasmid, the early simian virus (SV)40 promoter controlled the neomycin phosphotransferase gene that facilitated the selection of transgenic cells. Stability of methylation status and expression of the CAT gene were assessed in a number of clonal transgenic cell lines (Müller and Doerfler 1987).

The foreign DNA was integrated frequently in multiple tandems of the transfected plasmid. Among 19 clonal cell lines, the unmethylated construct remained in that state, and in 18 of these lines the CAT gene was continuously expressed. Among 14 cell lines transgenic for the premethylated construct, 7 lines failed to express the test transgene, and the three 5'-CCGG-3' sites in its late E2A promoter remained almost completely methylated. In 5 cell lines the promoter remained partly methylated and the CAT gene was only weakly expressed. In 2 cell lines, the premethylated promoter lost the 5m-C modification altogether, and the CAT gene was strongly expressed (Müller and Doerfler 1987).

## Factors Determining De Novo Methylation: Reinsertion of a Mouse Gene into Its Authentic Position

We further pursued the general question of which factors would affect the de novo DNA methylation in mammalian genomes. The mouse B lymphocyte tyrosine kinase (BLK) gene was re-integrated by homologous recombination into the genome of mouse embryonal stem (ES) cells. Two different plasmid constructs containing that gene were used for these experiments. One construct also carried the weak E2A late promoter of Ad2 DNA in front of the luciferase gene. In the second, this gene was controlled by the strong early SV40 promoter. Upon homing through homologous recombination to the authentic chromosome 14 or by heterologous recombination to many different loci in the mouse genome, methylation patterns in the integrates were assessed by restriction with the methylation-sensitive endonuclease HpaII or the insensitive MspI. The mouse BLK gene reinserted into the genome by homologous recombination had reestablished the identical methylation pattern characteristic for the authentic, non-manipulated mouse BLK gene (Hertz et al. 1999).

The extent of de novo methylation in the DNA segments adjacent to the BLK gene in the integrated construct depended on the promoter present in the plasmid construct and on the location of the recombined construct in the ES genome. In homologously inserted DNA, which carried the weak Ad2 promoter, de novo methylation was extensive. Presence of the strong SV40 promoter led to hypomethylation or no methylation at all. When the enhancer sequence was removed from the SV40 promoter, it also became hy-permethylated. All randomly integrated constructs, independent of the type of promoter or enhancer included, were hypermethylatedin patterns different from the original methylation pattern of the mouse BLK gene. We concluded (Hertz et al. 1999):

1. That an authentic mouse gene reinserted into its original genomic site was remethylated to the identical pattern as previously present on the target and on the allelic site
2. That heterologous recombination to randomlytargeted loci did not confer the mouse BLK gene-specific methylation pattern
3. That promoter strength in a construct was able to influence the pattern of methylation imposed de novo on the inserted construct after homologous recombination in the mouse genome

## De Novo DNA Methylation: An Ancient Cellular Defense Mechanism?

The de novo methylation of integrated foreign DNA is a phenomenon widely documented throughout the phyla of eukaryotic organisms, in mammalians as well as in plants (for review, see Meyer 1995). The currently established genomes have evolved over many millennia. We tend to assume that the evolution of genomes is a continuous process that has started right after the beginning of organismic life and probably before organisms arose and will continue as long as this biological system can be maintained. Even under experimental conditions, many organisms have been shown to be capable of accepting and accommodating foreign DNA, although with unpredictable—

advantageous or catastrophic—consequences for the acceptor cell. Since all cells in culture or organisms in environment-challenged life are subject to stringent conditions of selection, even the cell that has been forced by its innate recombination mechanisms to tolerate the genomic insertion of foreign DNA can avail itself of an ancient defense mechanism against the genetic activity of foreign DNA that could carry active genes.

Since promoter methylation has been identified as part of a mechanism for the long-term silencing of genes and DNA segments, the de novo methylation of integrated foreign DNA can be contemplated as such a defense mechanism or at least as an integral part of it (Doerfler 1991; Yoder et al. 1997). A large part of 5-mC residues is found in the parasitic sequence elements of retrotransposons and endogenous retroviruses that constitute more than 35% of the human genome. Perhaps intragenomic parasites are recognized by their high copy number. Long-term inactivation by DNA methylation also entails the possibility of 5-mC residues being deaminated to Ts followed by permanent inactivation (Bestor 1998).

## Are Integrated Foreign DNA Sequences Stabilized by Hypermethylation?

The Ad12-transformed cell line T637 was obtained by the in vitro transformation of BHK21 hamster cells with Ad12 (Strohl et al. 1967) and carries about 15 copies of Ad12 DNA integrated at a single chromosomal locus (Stabel et al 1980; Knoblauch et al. 1996; Schröer et al. 1997). Unintegrated, free Ad12 DNA could never be detected in any of the adenovirus-transformed cell lines or in Ad12-induced tumor cells. The integrated Ad12 DNA is methylated in functionally significant patterns (Sutter and Doerfler 1980; Hohlweg et al. 2003). Upon continuous propagation in cell culture, a small number of morphological revertants arose from the cell line T637 (Groneberg et al. 1978).

These revertants exhibitedafibroblastic phenotype in contrast to the more rounded, epithelioid appearance of the T637 cells. The revertants can be selected due to their resilience to spent, acidic medium in which T637 cells detach. We have studied several of these revertants with respect to their content of residual viral DNA (Eick et al. 1980).

In one of these revertants, TR12, only one complete copy of Ad12 DNA and an approximately 3.9-kb-long fragment from the right terminus of Ad12 DNA persist in the integrated state (N. Hochstein, I. Muiznieks, and W. Doerfler, manuscript in preparation). Studies using methylation-sensitive restriction endonucleases revealed that the integrated Ad12 DNA in the revertant cell line TR12 was even more extensively methylated than in the cell line T637 from which TR12 originated (Orend et al. 1995b). Preliminary results adduced from experiments using the bisulfite sequencing method confirm these data (N. Hochstein and W. Doerfler, unpublished). We propose that hyper- or nearly completely 5'-CG-3' methylated foreign DNA sequences would be more stably integrated than less completely methylated foreign genomes. Repetitive sequences, like endogenous retrovi-ral retrotransposons, appear to be significantly but not completely methylated (Heller et al. 1995).

## De Novo Methylation of Ad12 and of Cellular DNA in Hamster Tumors

The insertion of foreign DNA into established mammalian genomes has consequences for the inserted foreign DNA and for the recipient host genome. We have chosen to study these events in Ad12-induced hamster tumor cells, a model in basic research as well as for its significance in viral oncology. When Ad12 is injected subcutaneously into newborn hamsters within 24 h after birth, undifferentiated tumors develop at the site of topical application of the virus within a few weeks in 70%-90% of the animals that survive injection. As mentioned above, upon the intramuscular injection of Ad12 into the gluteal region, numerous tumors are found intraperitoneally Histologi-cally, these tumors exhibit Homer-Wright rosette-like structures indicative of primitive neuroectodermal tumors (PNET).

The tumor cells express proteins that are characteristic both for neu-roepithelial as well as for mesenchymal cells (Hohlweg et al. 2003, 2004). Each tumor cell carries multiple copies of integrated Ad12 DNA. Free viral DNA has never been found in any of these tumor cells. The cells from an individual tumor carry the viral integrates, with few exceptions, all at a single chromosomal location that is identical in all cells of a given tumor. Among 60 different tumors investigated, only one showed two different chromosomal loci to be occupied by Ad12 DNA when investigated by the fluorescent in situ hybridization (FISH) technique.

When we compared the sites of Ad12 DNA integration in more than 100 different hamster tumors, the sites of viral DNA integration in the host cell genome were different from tumor to tumor (Doerfler 1982; Kuhlmann and Doerfler 1982; Hilger-Eversheim and Doerfler 1997). We, therefore, favor a model of clonal origin of these Ad12-induced tumors. Aside from the de novo methylation of the viral integrates in all of the Ad12-induced tumors, there can be changes in the extent of DNA methylation also in the host genome (Heller et al. 1995). Moreover, the transcription patterns of cellular genes in different tumors can be very similar but there are also differences. The patterns of transcription of the integrated Ad12 genes are strikingly similar in all Ad12-induced hamster tumors analyzed and resemble those patterns in the Ad12-transformed hamster cell line T637 (Hohlweg et al. 2003, 2004).

## Loss ofAd12 Genomes IsCompatible with Maintenanceof the Oncogenic Phenotype

Upon continuous cell culture, some of the Ad12-induced hamster tumor cells, which carried multiple copies of integrated and de novo methylated Ad12 genomes, lost the viral DNA sequences completely or almost completely. Surprisingly, these revertants devoid of Ad12 DNA and notably lacking its left terminus with the transforming E1 region of the viral genome retained their oncogenic phenotypes when reinjected into hamsters (Kuhlmann et al. 1982; Pfeffer et al. 1999). We consider this result highly significant in that it demonstrates that Ad12 is capable of inducing tumors, which keep their oncogenic properties in animals, even when the E1 region is completely lost from these tumor cells. Many researchers in the field of adenovirus tumor biology consider the E1 region of the adenoviral genome akin to an oncogene and its maintenance paramount in preserving the oncogenic potential of the adenovirus-induced tumor cells.

However, this notion has been experimentally derived mainly by using rat embryo fibroblasts that were transformed in culture through the transfection of adenovirus type 5 (Ad5) DNA fragments (for review, see Zantema and van der Eb 1995). Obviously, tumor induction by Ad12 in animals can be a quite different process and can probably not appropriately be mimicked by transfection experiments in cell culture. In this view, global changes in the cellular genome as a consequence of viral DNA integration are invoked as an important part of the mechanism of viral oncogenesis in addition to the expression of a single viral "oncogene" (Doerfler 2000; Doerfler et al. 2001).

## DNA Methylation in non-CpG Dinucleotides; Hemimethylated DNA

The establishment of de novo patterns of DNA methylation in mammalian genomes is characterized by the gradual spreading of methylation, which has been documented to occur across multiple copies of integrated adenovirus genomes as well as, at the nucleotide level, in the integrated E2A promoter of Ad2 DNA (Müller and Doerfler 1987; Toth et al. 1989). A few 5'-CG-3' sequences can remain hemimethylated for several cell generations before they become totally methylated. Hemimethylation maybe a transient phenomenon but could also persist for a certain period in specific segments of a transgene or in the genome in general. In the Ad2-transformed cell line HE2, the E2A promoter in the integrated Ad2 genome is heavily methylated not only in all 5'-CG-3' dinucleotides but also in some 5'-CA-3' and5'-CT-3' dinucleotides (Toth etal. 1990). Evidence for the occurrence of 5-mCinnon-5'-CG-3' dinucleotides has also been presented by Woodcock et al. (1987). There is at present no plausible explanation concerning whether and how this type of methylation can be maintained following DNA replication.

## Initiation and Spreading of De Novo Methylation

The mechanism of de novo methylation is not well understood. From the results of experiments performed with the adenovirus system, the following conclusions appear well founded:

1. In a transgene the size of Ad12 DNA, 34,125-bp (Sprengel et al. 1994), de novo methylation starts in internal regions of the genome (Orend et al. 1991,1995a) and spreads from there

across the transgene (Toth et al. 1989,1990). In transformed or tumor cells, the early regions, particularly E1, can be partly spared from de novo methylation, because their gene products are required during the selection for the oncogenic phenotype.

2. Initiation of de novo methylation is regional and not confined to one or a few contiguous 5'-CG-3' dinucleotides (Hohlweg et al. 2003). In Ad12-induced tumors the extent of Ad12 transgene methylation is not only dependent on tumor size.
3. Transcribed regions of the transgene are hypomethylated, and inactive segments are hypermethylated (Sutter and Doerfler 1980; Muiznieks and Doerfler 1994a; Munnes and Doerfler 1997).
4. Spreading of DNA methylation is not entirely contiguous. For unknown reasons, certain 5'-CG-3' dinucleotides can remain unmethylated (N. Hochstein, I. Muiznieks, and W. Doerfler, manuscript in preparation).
5. The extent of de novo methylation and the speed of its spreading seem to be determined by the chromosomal site of transgene localization, and to a lesser extent by the nucleotide sequence of the transgene (Orend et al. 1995; Hertz et al. 1999).
6. The reestablishment of the authentic pattern of DNA methylation in the correctly reinserted BLK gene in mouse embryonic stem cells (Hertz et al. 1999) implies that each segment of the mammalian genome is capable of exerting a memory function that could be directly related to the mechanism of maintenance methylation. Specific chromatin structures at different sites of the genome may be decisive in directing the methylation reaction. In sequences adjoined to the BLK gene, the de novo methylation patterns depended on the strength of the viral promoter in the construct. The weaker E2A late promoter of Ad2 DNA led to a more extensive de novo methylation than the presence of the stronger early promoter of SV40 DNA.
7. In the same set of experiments (Hertz et al. 1999), we noted that the same re-integrated BLK gene, when it was inserted at

randomly selected sites in the genome, was hypermethylated in patterns completely different from the original BLK gene pattern and independently of promoter strength.

8. When Ad12-transformed cells or Ad12-induced hamster tumor cells were continuously passaged in culture, revertants arose that had lost all or a part of the multiple copies of integrated Ad12 DNA. The then-persisting Ad12 DNA genomes in the revertants seemed to be more completely methylated than the lost copies. Hence, the idea was put forward that the levels of DNA methylation of transgenes might be related to their stability of fixation in the host genome (Orend et al. 1995b). Moreover, continuous culture of different Ad12-induced tumor cells led sometimes to the selection of cell lines with very similar integration patterns (Orend et al. 1994). It is conceivable that the integration sites in the cells selected this way were most compatible with the survival of these cells in culture.

## INVERSE CORRELATIONS BETWEEN PROMOTER ACTIVITY AND METHYLATION

ThefieldofDNA methylationinmammalian cells was pioneeredbythediscov-ery of inverse correlations between the extent of segmental DNA methylation and the genetic activity of these segments. In integrated Ad12 DNA in Ad12-transformed cells, the early viral genes are transcribed, and the late Ad12 genes responsible for the synthesis of virion capsid proteins are permanently silenced. Hence, in Ad12-transformed cells or in Ad12-induced tumor cells, late viral gene products and mature virions are not synthesized (Ortin et al. 1976). Thus, Ad12-transformed cells provided a suitable tool to study the levels of DNA methylation in distinct sections of the viral genome and to document inverse correlations to gene transcription for the first time (Sutter and Doerfler 1979, 1980; Vardimon et al. 1980, 1981).

A particularly clear example has been offered by Ad2-transformed cell lines HE1, HE2, and HE3 (Cook and Lewis 1979). The E2A region of Ad2 DNA is not expressed in cell lines HE2 and HE3, whereas cell line HE1 does express the E2A region of the integrated Ad2 DNA (Johansson et al. 1978). Accordingly, the 5'-CCGG-3' dinucleotides

(HpaII sites) in the promoter region of this gene are unmethy-lated in cell line HE1 but methylated in cell lines HE2 and HE3 (Vardimon et al. 1980).

At the time, one had to rely on the analyses of a limited number of 5CG-3' dinucleotides whose state of methylation could be assessed only by the use of methylation-sensitive restriction endonucleases, in this example of HpaII and MspI (5'-CCGG-3').

These seminal observations started a burst of similar studies with numerous viral and cellular genes and confirmed almost without exception the initial observations that suggested that specific promoter methylation patterns are instrumental in the long-term silencing of eukaryotic genes (for review, see Doerfler 1983; Munnes and Doerfler 1997). To this day, however, we do not understand whether there have to be a specific number or pattern of 5-mC residues in a promoter to assure its long-term silencing.

## Productive Versus Abortive Infection of Cells with Ad1 2

Ad12 interacts with human cells in a productive infection leading to the synthesis of a large number of progeny virions. In contrast, the replication of Ad12 in hamster cells is completely blocked (Doerfler 1969, 1991), possibly because only few copies of Ad12 DNA are capable of reaching the nucleus of the hamster cells (D. Webb, M. Hösel, B. Schmitz, et al., submitted).

However, the expression of the Ad 12 genome in hamster cells appears to be suppressed in several steps of the normal replication cycle. Nevertheless, viral DNA can be traced in the nucleus of the abortively infected hamster cells and limited transcription of the early genes of non-integrated Ad12 has been documented (Ortin et al. 1976; D. Webb, M. Hösel, B. Schmitz, et al., submitted).

There is, however, no methylation of the free intranuclear Ad12 DNA in productively or abortively infected cells (Vardimon et al. 1980). Obviously, methylation of free viral DNA, even in the abortive system, does not serve to regulate or to inactivate the late Ad12 viral genes that are not transcribed in abortively infected hamster cells.

## The Actively Transcribed Genome of Frog Virus 3 Is Completely 5-CG-3-Methylated

The hypermethylated state of the virion-encapsidated or of the intracellular FV3 genomes (Willis and Granoff 1980) in fish or mammalian cells has taught us that the biological significance of DNA methylation cannot beschematically interpreted and depends entirely on the biological system studied. While the inverse correlations described above hold true for most systems investigated so far, there are notable exceptions to this "rule" and, of course, no such stringent dogmata in biology. It has been documented that theviral L1140 gene is actively transcribed late after infection offish cells with FV3, although it is methylated in all 5-CG-3' dinucleotides (Munnes et al. 1995).

In FV3-infected fish or hamster cells, a transfected L1140 promoter-indicator gene construct is active in the unmethylated or fully 5'-CG-3'-methylated form. When the same construct is methylated only in the 5'-CCGG-3' (HpaII) sequences, its activity is reduced. Compatibility of the methylation of an immediate-early FV3 promoter with its active transcription has also been reported (Thompson et al. 1988). These data confirm the special methylation requirements of this promoter in FV3 DNA. Specialproperties of the FV3 DNA-protein interactions may account for these unexpected activity patterns. It would be interesting to study in greater biochemical detail the transcription of FV3 genes and the proteins involved in their regulation.

## Site-Specific Promoter Methylation and Gene Silencing

The finding of inverse correlations between promoter activity and extent of DNA methylation led to the concept that sequence-specific promoter methy-lations exert a regulatory function on gene activity. In order to provide more direct evidence for this interpretation, we devised experiments in which a number of promoter-indicator gene constructs were tested for their genetic activities in the unmethylated or in the methylated state at 48 h after transfec-tion into mammalian cells in culture. In general, these data corroborated the earlier interpretation of promoter inactivation by promoter methylation, although this experimental approach could, of course, not help decide whether in an intact mammalian genome promoter methylation was the cause or consequence of promoter inactivation. The former possibility, however, remains the more likely explanation.

## The E2A Promoter of Ad2 DNA

In a first set of experiments, the oocyte system from *Xenopus laevis* was adapted to test unmethylated or methylated promoter-gene constructs for genetic activities. The cloned E2A region of Ad2 DNA was then 5'-CCGG-3'-methylated with the HpaII DMTases or was left unmethylated. Subsequently, either construct was microinjected into the nuclei of *X laevis* oocytes. The methylation status of these constructs was maintained in the oocyte nuclei.

At 48 h after microinjection, the unmethylated construct was transcribed in the oocyte nuclei, while the methylated construct was silenced (Vardimon et al. 1982a). Transcription was initiated at the authentic E2A late promoter of Ad2 DNA. Control constructs carrying the unmethylated histone h22 gene were actively transcribed when co-injected with the methylated and silenced E2A construct. Hence, there was no evidence for possible unspecific inhibitory effects exerted by the in vitro premethylated construct. Modification of the E2A construct by the BsuRI (5'-GG*CC-3') DNMT did not inactivate transcription (Vardimon et al. 1982b). These data provided direct evidence for the notion that 5'-CG-3' sequence-specific promoter methylation was involved in the silencing of eukaryotic genes.

The system was further refined by separating the promoter of the E2A gene from its body and by preparing both DNA fragments in quantitative amounts. We then 5'-CCGG-3'-methylated either the promoter or the gene body part of the constructs. Subsequently, the methylated promoter was re-ligated to the unmethylated body of the E2A gene and, conversely, the unmethylated promoter was reattached to the methylated E2A gene sequence. Upon microinjection into the nuclei of *X laevis* oocytes, only the construct, in which the promoter had been methylated, was inactivated. The construct with an unmethylated promoter but a methylated gene body was actively transcribed (Langner et al. 1984). We interpreted these data to demonstrate that sequence-specific promoter methylation led to gene inactivation, whereas the methylation of the body of this gene did not affect its activity.

## The El A Promoter of Ad12 DNA

A similar set of experiments was performed with constructs that carried the chloramphenicol acetyltransferase (CAT) gene as an indicator for gene activity under the control of the E1A regulatory region of Ad12 DNA. Methylation of the two HpaII (5'-CCGG-3') or of the seven HhaI (5'-GCGC-3') sequences in this promoter inactivated the CAT gene or severely decreased its activity at 48 h after the transfection of these constructs into mouse Ltk- cells (Kruczek and Doerfler 1982,1983). Several additional sites in the promoter of the E1A gene of Ad12 were methylated, and the activity of the modified promoter was assessed with the CAT indicator gene.

The C-residue methyla-tion of two AluI sites (5'-AGCT-3') downstream from the TATA box had no effect on promoter activity. However, when one EcoRI (5'-GAATTC-3') sequence, 281 bp upstream, or one TaqI (5'-TCGA-3') site downstream from the TATA signal in the promoter was deoxyadenosine methylated, the promoter became silent (Knebel and Doerfler 1986).

Deoxyadenosine methylation of an MboI (5'-GATC-3') sequence downstream of the TATA signal had no effect. Apparently, methylated nucleotides introduced at highly specific promoter locations can play an important role in the downregulation of the Ad12 E1A promoter at least in transfection experiments. Since $N^6$-mA is not known to occur in mammalian DNA, the effect of $N^6$-mA on promoter activity has been unexpected.

In an extension of this experimental approach, additional viral and non-viral eukaryotic promoters were tested for their sensitivity towards 5'-CG-3' or 5'-CCGG-3' methylation. The CAT or luciferase gene was used as activity indicator 24 h after the transfection into different human cell lines (HeLa, PA-1, 293). The methylation of all 5'-CG-3' sequences by the SssI DNMT inactivated the E2A late promoter of Ad2 DNA, the human cytomegalovirus promoter, the TNF-a promoter, the herpes simplex virus thymidine kinase promoter, and decreased the activity of the SV40 early promoter (Muiznieks and Doerfler 1994a). In some experiments, HpaII methylation just led to a decrease in the genetic activity of some of these constructs.

## The L1140 Promoter of Frog Virus FV3 DNA

The resistance of the late L1140 or an early FV3 promoter to complete 5'-CG-3' methylation and its full activity in fish or mammalian cells in the completely methylated state has been described (Thompson et al. 1988; Munnes et al. 1995).

## The p10 Promoter of the AcNPV Insect Virus

A construct, which contained the promoter of the p10 gene of the insect virus *Autographa californica* nuclear polyhedrosis virus (AcNPV) and the CAT indicator gene, was active in AcNPV-infected *Spodoptera frugiperda* insect cells at 18 h after transfection of the construct.

When the three 5'-CCGG-3' (HpaII) sites in the promoter and its downstream region were methylated, the p10 gene promoter was silenced (Knebel et al. 1985). Although insect cells may contain only minor amounts of 5-mC, the activity of an AcNPV insect virus promoter could be shown to be sensitive to sequence-specific methylation.

## Human Alu Sequences Transcribed by RNA Polymerase III

We have also demonstrated that the polymerase III transcription of Alu sequences associated with the human angiogenin, the tissue plasminogen activator (tPA), or the ot1-globin gene is inhibited by 5'-CG-3' methylation of these sequences (Kochanek et al. 1993).

Their methylation also interferes with the binding of proteins to the B control region of these Alu sequences (Kochanek et al. 1995).

## Bending of Promoter DNA Sequences Due to Methylation?

The site-specific methylation in 5'-CCGG-3' (HpaII), 5'-CGCG-3' (FnuDII), or in 5'-CG-3' (SssI) sequences of the E2A promoter, the polymerase III-transcribed virus-associated RNA I (VAI) gene of Ad2 DNA or of the human angiogenin gene-associated Alu sequence can alter the electrophoretic mobility of these DNA sequences in non-denaturing polyacrylamide gels.

This finding indicates that the bending of the tested sequences might be altered by DNA methylation (Muiznieks and Doerfler 1994b).

## An Adenovirus E1A Gene Product or the Strong Enhancer of Human Cytomegalovirus Can Overcome the Transcription-Inactivating Effect of Promoter Methylation

The removal of the methyl group from 5-mC in a methylated promoter in the absence of DNA replication seems to be a rare event. Hence, other mechanisms for transient reactivation of a permanently methylated promoter appear to be required. Of course, experimentally, the methylated E2A late promoter in the Ad2-transformed cell line HE3 can be demethylated and reactivated by growing the cells in culture in the presence of 50 μM 5-azacytidine (5-aza-C), an inhibitor of maintenance methylation (Knust et al. 1989). This approach provides support of principle but does not adequately mimic the situation in a biological system.

In human 293 cells, which carry the left terminus of Ad5 DNA chromoso-mally integrated and express the E1 region of Ad5 constitutively, the inactivating effect of 5'-CCGG-3' methylation of an E2A promoter construct of Ad2 DNA is released or markedly decreased (Langner et al. 1986). We have also shown that the E1A gene encoding the 13S RNA and the 289-amino acid (aa) protein of Ad2, a well-known transactivator of genes (Flint and Shenk 1989; Nevins et al. 1995), is responsible for the reversal of the inactivating effect of E2A promoter methylation (Weisshaar et al. 1988). It is unknown by which mechanism the 289-aa E1A function is capable of effecting this reactivation. The methylated E2A promoter did not lose its 5'-CCGG-3' methyl groups in the reactivation process at 48 h after transfection. Moreover, the authentic cap site of this promoter was used in the transcription following reactivation (Weisshaar et al. 1988; Knust et al. 1989).

Similarly, the 5'-CCGG-3' methylated E2A promoter of Ad2 DNA was active when the strong immediate early enhancer of HCMV DNA was inserted into the promoter-indicator gene construct in a position either immediately antecedent to the promoter or several thousand nucleotides remote from it (Knebel-Mörsdorf et al. 1988). Transcription was initiated correctly at the authentic cap site of the E2A gene, and 5'-CCGG-3' methylation remained unaltered at least during the duration of the transient expression experiment.

## Promoter Methylation and Protein Binding

This topic has been extensively investigated in several laboratories (addressed in many of the chapters of this volume). In the E2A promoter system of Ad2 DNA, the in vitro methylation of 5'-CCGG-3'sequences at nucleotides +24, +6, and-215 relative to nucleotide +1, the site of transcriptional initiation, was demonstrated to lead to transcriptional inactivation in transient expression studies inX *laevis* oocytes (Langner et al. 1986), in mammalian cells (Langner et al. 1986), after the genomic fixation of the promoter in mammalian cells (Müller and Doerfler 1987), and in a cell-free transcription system using nuclear extracts from human HeLa cells (Dobrzanski et al. 1988). DNA fragments 50 or 73 bp in length—which comprise the +24 and +6 5'-CCGG-3' sequences of the E2A promoter of Ad2 DNA in the unmethylated, methylated, or hemimethylated state—were incubated with partly purified nuclear extracts from human HeLa cells.

Protein binding to these DNA preparations was assessed by electrophoretic mobility shift assays (EMSAs). The formation of one of the observed DNA-protein complexes in this system was compromised when the construct was methylated or hemimethylated (Hermann et al. 1989). The results of the necessary competition experiments confirmed the interpretation that specific promoter methylation interfered with the binding of nuclear proteins from human cells. There was evidence that the AP2 transcription factor was among the proteins sensitive to promoter methylation in this system (Hermann and Doerfler 1991).

## PATTERNS OF DNA METHYLATION IN THE HUMAN GENOME

A more profound understanding of the multifaceted biological functions of DNA methylation in mammalian and other genomes will remain elusive unless we have at hand the complete nucleotide sequences of these genomes including the fifth nucleotide. Researchers interested in the function of 5-mC have, therefore, been disappointed by the otherwise admirable results of the Human Genome Project.

The human epigenome project has been initiated, and its results, once at least partly completed, will undoubtedly fill a serious gap in the anatomy of the human genome. In the early 1990s, my laboratory began, as a pilot project as it were, to study DNA methylation patterns in various parts of the human genome. A part of these results has been summarized (Doerfler 2000). A more complete survey will be presented here. Our studies also had the aim of contributing to the understanding of epigenetic mechanisms and of human disease.

## Interindividual Concordance in Human DNA Methylation Patterns

We have asked the question of how tightly preserved patterns of DNA methylation actually are in the promoter and 5'-upstream regions of a human gene among several individuals of different ethnic origins.

The human genome, like many other eukaryotic genomes, is characterized by the existence of complex patterns of DNA methylation that reflect, in an unidentified way, states of gene activity and inactivity and, equally important and related, the chromosomal structure of the (human) genome. The 5'-upstream and promoter regions of the human genes for TNF-a and TNF-p were investigated with the bisulfite sequencing technique (Frommer et al. 1992; Clark et al. 1994) for the presence of 5-mC residues (Kochanek et al. 1990, 1991). Human DNA was derived from peripheral blood gran-ulocytes, lymphocytes, or from sperm. The results indicated that patterns of DNA methylation, at least in these genome segments, were interindivid-ually highly conserved.

Thus, in the TNF-a DNA from granulocytes of 15 individuals of African, Caucasian, or Chinese origin, the 5-mC residues were consistently found in 5-CG-3' dinucleotide positions minus IX, minus X, minus XI (upstream of the cap site), and in position plus XVI (downstream of it). Very different distributions of 5-mC residues were observed in human cell lines HL60, Jurkat, and RPMI1788. The TNF-a gene is transcribed in human granulocytes.

A very different result emerged for the promoter and upstream regions of the human gene for TNF-p that is not transcribed in human granulocytes. All 13 5'-CG-3' dinucleotides in this segment were methylated, two only hemimethylated. Again, this pattern held true in the granulocytes from nine different individuals. The same sequence was completely unmethylated in human lymphocytes from the same individuals, in sperm and in the human cell lines RPMI 1788 and HL60, but almost completely methylated in cell line Jurkat (Kochanek et al. 1990).

These data document that methylation patterns in human DNA can be very different in different cell lines, but can be interindividually highly concordant. Moreover, patterns of DNA methylation in specific genome segments can vary a great deal.

The patterns of DNA methylation in the human TNF-a and TNF-p genes in granulocytes, monocytes and in several cases of acute (AML) or chronic myeloid leukemia (CML) were found to be very similar, except for one AML, in which the region in the TNF-a gene was completely unmethylated, and several leukemia cases in which many sites in the TNF-p gene were only hemimethylated (Kochanek et al. 1991).

In T and B lymphocytes of many individuals and in a number of Hodgkin and non-Hodgkin lymphomas, both the TNF-a and TNF-p genes were un- or hypomethylated. The DNA in HeLa cells in culture was completely methylated in the upstream and promoter regions of both genes (Kochanek et al. 1991). If leukemia- or lymphoma-specific patterns should exist, they are very complex and not readily recognizable by this type of analysis.

We have also compared methylation patterns by HpaII (5'-CCGG-3') and HhaI (5'-GCGC-3') cleavages of human DNA from European and Japanese individuals across about 500 kb of randomly selected DNA sequences in the human genome and found complete interindividual congruence of patterns by this method of admittedly intermediate sensitivity (Behn-Krappa et al. 1991).

## Methylation Patterns in Genetically Imprinted Regions of the Human Genome

The Prader-Willi/Angelman region on chromosome 15q11-q13 of the human genome is genetically imprinted, i.e., on the maternally and on the paternally inherited chromosome different genes are activated and others silenced. The molecular mechanisms underlying genetic imprinting are not completely understood. However, there is much evidence that differences in the methylation patterns in imprinted regions on the two alleles play an important role in the imprinting phenomenon. As part of a study on DNA methylation patterns in the human genome, we investigated all 5'-CG-3' dinucleotides in the vicinity of exon 1 of the SNRPN and the D15S63 loci on chromosome 15q.

The SNRPN transcripts might be involved in imprint switching during gametoge-nesis. By using the bisulfite protocol of the genomic sequencing technique, we looked at individual chromosomal PCR products from normal individuals, from Prader-Willi and from Angelman patients. In this region, non-5'-CG-3' C-residues were never methylated. Around exon 1 of the SNRPN gene, more than 96% of the 23 5'-CG-3'-dinucleotides were methylated on the maternal chromosome, as apparent from the genomic sequencing data with DNA from Prader-Willi patients in whom this segment was deleted on the paternal chromosome. In contrast, the same region on the paternal chromosome was completely devoid of methylated 5'-CG-3' dinucleotides (Zeschnigk et al. 1997a).

Angelman syndrome patients carry a deletion of the region on the maternal chromosome. The methylation status in the D15S63 locus, however, was quite different in that only two CfoI/HhaI sites were methylated, compared with more than 96% on the maternal chromosome. The remaining five 5'-CG-3' dinucleotides in this segment were methylated only 45%-70% on the maternal, and only 5%-14% on the paternal chromosome (Zeschnigk et al. 1997a, b).

In an extension of this study (Schumacher et al. 1998), it was demonstrated again by bisulfite genomic sequencing that the 16 5'-CG-3' dinucleotides in the 1.15 kb AS-SRO region on human chromosome 15q were methylated to 83%-87% on both the maternal and paternal chromosomes in healthy individuals as well as in Prader-Willi and Angelman syndrome patients.

There maybe alow degree of mosaicism but there are no parent-of-origin-specific differences in the methylation patterns in this part of the genome (Schumacher et al. 1998). These findings attest to the significant variability of the methylation patterns even in imprinted parts of the human genome.

## PATTERNS OF 5-CG-3 METHYLATION IN THE PROMOTER OF THE FMR1 GENE: RELEVANCE FOR THE FRAGILE X SYNDROME

In patients suffering from fragile X (FRAXA) syndrome, a naturally occurring 5'-$(CGG)_n$-3' repeat in the promoter and the 5'-untranslated regions (5'-UTR) of the FMR1 gene on human chromosome Xq27.3 is expanded excessively.

In normal individuals, the value *n* ranges from 6 to 40, in pre-mutation females *n* assumes values between 40 and 200, while in affected individuals the repeat *n* lies over 200 and can exceed 2,000. The expanded repeat is hypermethylated, perhaps because such expansions are recognized as foreign DNA and become subject to modification. The ensuing inactivation of the FMR1 gene is the most likely cause for the disturbed embryonal and fetal development that is the basis for the syndrome (for review, see Sutherland et al. 2002).

By applying the bisulfite genomic sequencing technique, we determined the methylation profiles in the promoter and 5'-UTR of the FMR1 gene on single chromosomes of healthy individuals and of selected premutation carriers and FRAXA patients (Genç et al. 2000). In the DNA from FRAXA patients, there is considerable variability in the lengths of the 5'-$(CGG)_n$-3' repeats and in the levels of methylation in the repeats and the 5'-UTR regions in that all patients seem to be mosaics with respect to both parameters. In one patient with repeat lengths between *n* = 15 and greater than 200, shorter repeats *(n* = 20-80) were methylated or unmethylated, longer repeats *(n* = 100-150) were often completely methylated.

A particular repeat in this patient with *n* = 160 proved to be completely devoid of 5-mC residues. This repeat mosaicism was observed in several FRAXA patients analyzed in our laboratory (Genç et al. 2000). As expected for healthy females with one at least partially inactivated X chromosome, hypermethylated repeats and5'-UTR sequences were found.We also demonstrated that theauthentic FMR1 promoter from healthy individuals was sensitive to methylation as shown by comparing the transient activities of constructs carrying the luciferase gene under the control of the unmethylated or the SssI (5'CG-3') completely methylated FMR1 promoter in human HeLa or 293 cells (Genç et al. 2000). The methylated, inactive FMR1 promoter regions did not bind to specific cellular proteins as determined by footprinting analyses, whereas active, unmethylated promoter regions did bind proteins (Schwemmle et al. 1997).

## The Promoter and 5-Upstream Region of the RET Protooncogene: A Gene Involved in the Causation of Hirschsprung Disease

The RET (rearranged during transformation) protooncogene plays a role in the causation of some familial or sporadic cases of Hirschsprung disease that results from an impaired development of the neural crest-derived neurons of the enteric ganglia (for review, see Passarge 2002). We investigated the level of DNA methylation in a DNA segment of about 1,000 bp in the promoter and 5'-upstream region of this gene (Munnes et al. 1998). By again applying the bisulfite genomic sequencing technique, DNA from peripheral white blood cells (PWBCs) from healthy individuals or from Hirschsprung disease patients was used as well as DNA from different human tissues and from a human embryonic kidney cell line. In a DNA section starting about 790 bp upstream from the transcriptional start site, a few 5-mC residues were found. However, in a 5'-CG-3' rich 400-bp stretch in the RET gene promoter with 49 such dinucleotide pairs, not a single 5-mC residue was present, although the RET protooncogene was not transcribed in many human tissues.

Weak transcriptional activity was detected in many neural crest-derived human tissues. Obviously, the RET gene promoter was not silenced by a long-term signal-like promoter methylation, possibly because it had to be expressed occasionally, and its transcription was controlled by factors other than DNA methylation. In in vitro experiments in which the transcriptional activity of the RET gene promoter was assessed in linkage to an indicator gene after transfection into human cells, the activity of this promoter was decreased by HpaII (5-CCGG-3') methylation and abolished by SssI (5'-CG-3') methylation. Hence the RET protooncogene promoter is sensitive to DNA methylation at least in transfection and transient transcription experiments (Munnes et al. 1998).

## Genes for Proteins in the Human Erythrocyte Membrane

Alterations in the structure of the erythrocyte membrane can be related to mutations in specific genes for proteins that are essential elements of this membrane. These structural alterations in the erythrocyte membrane are responsible for hematological diseases, like hereditary elliptocytosis or hereditary spherocytosis (for review, see Yawata 2003). By the bisulfite genomic sequencing procedure we determined patterns of methylation in the promoter and 5-regions of the following human genes: the protein 4.2 (P4.2) gene *(ELB42),* the band 3 (B3) gene *(EPB3),* the p-spectrin (P-Sp) gene *(SPTB),* and the ankyrin gene in several individuals (Remus et al. 2001, 2005). The promoter regions of the *EPB3* and *ELB42* genes were extensively methylated, whereas the promoters of the *SPTB* and the ankyrin genes were unmethy-lated. This finding again points to the interindividual conservation of certain patterns in the distribution of 5-mC residues in the human genome.

The human *SPTB* promoter conforms to expectations in that it is un-methylated and fully active throughout erythroid development. In contrast, high levels of promoter methylation correlate with promoter activity for the *EPB3* and *ELB42* genes during their sequential activation in erythrocyte differentiation (R. Remus et al., in press). In this respect, the *EPB3* and *ELB42* genes may resemble the genes of frog virus FV3.

This analysis was extended to patients with red cell membrane diseases, such as complete P4.2 deficiency due to mutations in the *ELB42* gene, with hereditary spherocytosis with *EPB3* mutations, and to hereditary elliptocyto-sis with mutations in the *SPTB* gene. Patterns of methylation in these patients were in general very similar to those of normal individuals (Remus et al. 2005).

## Promoter and Exon 1 of the Human Gene for the Interleukin 2-Receptor a Chain

The interleukin (IL)-2Ra gene is expressed in stimulated, but not in resting human T lymphocytes and plays an important role in promoting the T cell-mediated immune response. The-300 to +300 promoter/exon 1 region of the IL-2Ra gene was analyzedbythe genomic sequencing technique for its content of methylated 5'-CG-3' dinucleotides. In the cell types investigated—sperm, placenta, granulocytes, T-CLL, B-CLL,

Jurkat, KB cells—5-mC residues were not found in unstimulated or in stimulated lymphatic cells (Behn-Krappa and Doerfler 1993). The 5'-CG-3' sequence in position +198 was partly methylated. Even in cell types not relevant for the immune response, like in the human KB cell line, this regulatory region was consistently unmethylated. The promoter of a functionally essential human gene would not be long-term silenced by the methylation signal or else it could not be flexibly reactivated upon demand. Obviously, transient mechanisms of gene shut-off would be operationally preferred in these instances and thus remain independent of DNA methylation.

## Human Alu Sequences Associated with Specific Genes

The human Alu sequences that belong to the short interspersed repeat elements (SINE) comprise about 5% of the human genome and are about 300 bp long. The Alu elements might have been derived from reverse transcripts since they carry a 3'-dA-rich track. We analyzed the state of DNA methylation in the human Alu sequences associated with the genes for a1-globin, tissue plasminogen activator (tPA), adrenocorticotropic hormone (ACTH), and for angiogenin. DNA was investigated from lymphocytes, granulocytes, brain, heart muscle, and sperm as well as from human HeLa and KB cells. Both methylation-sensitive restriction enzymes and genomic sequencing techniques were employed.

In primary human cells, these Alu elements were highly methylated, but there were distinct differences in specific Alu sequence elements. In the DNA from haploid spermatozoa, Alu elements were often hypomethylated. The in vitro transcription of Alu elements was inhibited by 5'-CG-3' methylation of these sequences (Kochanek et al. 1993). The patterns observed in these specific Alu elements were identical in different individuals. The high level of DNA methylation in the Alu sequences associated with specific genes was consistent with their transcriptional silencing.

## Promoter of the Polymerase I-Transcribed Human Ribosomal Genes

The 5'-CG-3'-rich promoter region of the DNA-dependent RNA

polymerase I (rDNA) genes was analyzed for the presence of 5-mC nucleotides by the ge-nomic sequencing technique in DNA from primary human cells, from human tumor cells, and from human cell lines (Kochanek et al. 1996a). In none of the primary human cells or tumor cells was the rDNA promoter methylated. In contrast, in the DNA from the human cell lines, HeLa (cervical cancer), KB (oral cancer), Jurkat (T cells), or CEM (T cells) cells, the 5'-CG-3' dinu-cleotides were methylated between 50% (KB) and 85% (Jurkat). Apparently in the primary human cells, in granulocytes, T lymphocytes, and spermatozoa, as well as in chronic T cell (T-CCL), myeloid (CML), and B cell (B-CLL) leukemia cells, which are all actively dividing, the essential rDNA genes need be transcribed actively and are not methylated (Kochanek et al. 1996a). In cell lines, rDNA genes are also actively expressed, and alternate mechanisms of overexpression must exist. Could some of the rDNA genes be over-amplified?

## Randomly Selected Human Genes in Different Hodgkin's Lymphoma and Leukemia Cell Lines and in Normal Human Lymphocytes

Several of the protooncogenes, parts of the TNF-a and TNF-p genes, the insulin receptor, and lamin C genes were investigated by using the methylation-sensitive restriction enzyme HpaII and the control enzyme MspI. There were regions completely devoid of methylation, while others were completely or partly methylated. Various lymphoma and leukemia cell lines differed among each other in different regions of the genome and differed again from the patterns observed in normal primary human lymphocytes (Achten et al. 1991). Obviously, there is great variability, and no simple rules can be derived for the general characteristics of methylation patterns in leukemic versus normal human white cells.

## The Promoter of the Human 5-(CGG) $_n$-3-Binding Protein 1

From the nuclei of human HeLa cells, we isolated a 20-kDA protein that binds specifically to 5'-(CGG) $_n$-3' sequences (Deissler et al. 1996, 1997) and that might play a role in the control of promoters rich in CGG sequences like the FMR1 promoter in human DNA (Müller-Hartmann et al. 2000). The human gene for this protein, termed

CGGBP1, was located to human chromosome 3p, and its promoter was characterized in detail (Naumann et al. 2004). In several different human cell types, this promoter was unmethylated. The complete in vitro premethylation of all 18 5'-CG-3' dinucleotides in this promoter led to its inactivation upon transfection into human HeLa cells with the luciferase gene as activity indicator (Naumann et al. 2004).

## Towards a Complete Nucleotide Sequence of the Human Genome

Projects have been initiated to complete the human genomic DNA sequence by including the fifth nucleotide and to initiate a Human Epigenome Project (for review, see Beck and Olek 2003; Rakyan et al. 2004). Obviously, this will be a very important, but at the same time demanding, task. In Sect. 6 of this chapter, I have summarized our contributions, of course not towards the solution, but towards a more general appreciation of the problem that the mammalian genomes harbor functionally important patterns of DNA methylation. The distribution of 5-mC residues in the human genome is thought to be essential for the understanding of chromatin structure and of the regulation of human gene expression in the many different cell types and during development.

From the available data, the following list of problems and desirable approaches can be compiled. Of course, we want to be cautious and cannot claim general validity of any of the presently plausible observations and conclusions.

1. Patterns of 5-mC distribution across the human genome are highly specific. Each region, each promoter exhibits its own individual pattern.
2. The patterns can be interindividually conserved at least in several regions of the human genome.
3. It is likely that each human cell type could have a different pattern in each genome segment.
4. Long-term promoter inactivity is generally associated with hypermethy-lation of the promoter. Inactive promoters can, however, also be un- or hypomethylated, particularly when they

have to be occasionally reactivated. The state of promoter methylation by itself cannot reveal the activity status of a promoter.

5. Promoter strength might affect the pattern imposed upon a particular promoter.
6. There are distinct differences in the patterns of methylation between normal human lymphocytes and lymphoma or leukemia cell lines in different segments of the genome. It is at present not possible to derive functionally meaningful conclusions from these differences other than that there are extensive alterations. We pursue the possibility that the process of oncogenic transformation of a human cell is associated, possibly causally, with global changes in the genome organization that is also reflected in drastically altered methylation patterns. From our work on Ad12-induced hamster tumors and on Ad12 DNA-transgenic or bacteriophage ë DNA-transgenic hamster cells (see next section), we consider it likely that the insertion of foreign DNA is at least partly responsible for these alterations. Of course, it remains to be determined whether any of the global changes in methylation and transcription patterns demonstrated in tumor cells are the cause or the consequence of oncogenic transformation.

## Alterations of Cellular DNA Methylation upon Foreign DNA Insertion

In the Ad12-transformed hamster cell line T637 with 15 copies of viral DNA inserted at a single chromosomal site, extensive alterations, mainly increases, in the levels of DNA methylation in the HpaII (5'-CCGG-3') and HhaI (5'-GCGC-3') sequences were apparent in the retrotransposon sequences of the about 900 copies of intracisternal A particle (IAP)I genomes (Heller et al. 1995). The roughly 900 copies of IAPI sequences are a constitutive part of the hamster genome (Lueders and Kuff 1981; Ono and Ohishi 1983). In Ad12-transformed hamster cells, extensive changes in DNA methylation were also noted in the MHC class I and II, the Ig Cμ, the serine protease, and cytochrome P450 genes of the hamster cell genome. At least in the IAPI sequences, the increases in DNA methylation persisted in the revertant TR3 that had lost all 15 copies of Ad12 DNA (Heller et al. 1995).

Apparently, the alterations of the methylation patterns in the cellular genome are not dependent on the continued presence of the viral transgene DNA. The approximately 900 copies of IAPI DNA are distributed among many of the hamster chromosomes, often on their short arms (Kuff et al. 1986; Meyer zu Altenschildesche et al. 1996). Since the increases in IAPI methylation were extensive, the effects of the Ad12 DNA insertion at a single chromosomal site had to transgress this site and led to a disturbance in *trans* of DNA methylation patterns in the cellular genome, even on different chromosomes.

Repetitive sequences in the mammalian genome appear to be particularly prone to respond with altered methylation patterns to perturbations in the genome caused by foreign DNA insertions. We surmise that the selection of the genes and DNA segments influenced in *trans* by the foreign DNA integration event might depend on the site of transgene insertion. The mechanism of this modulation of DNA methylation in the recipient genome remains unknown; it might be sought in the direct interaction of neighboring chromosomes. Soluble factors could obviously also play a role.

Further evidence in support of the contributions that foreign DNA insertions rendered in altering DNA methylation patterns in the recipient genome came from experiments in which we generated clonal hamster BHK21 cell lines with multiple copies of integrated bacteriophage *X* DNA. The integration phenomena of *X* DNA resembled those of Ad12 DNA in that multiple copies of the phage DNA came to reside at a single site of the hamster chromosome and became progressively de novo methylated. However, in contrast to the integrated Ad12 DNA, the integrated *X* DNA was not detectably transcribed. Alterations in cellular DNA methylation patterns were also observed in the IAPI sequences and could be unequivocally documented by the bisulfite genomic sequencing method with which 35 5'-CG-3' dinucleotide positions were analyzed in a subsegment of the IAPI DNA region (Remus et al. 1999). Even a transcriptionally inert transgene, like *X* DNA, had led to alterations in the methylation profiles in the IAP transposons.

The question arose whether such differences in methylation patterns among the different copies of the roughly 900 IAPI equivalents might have preexisted in different BHK21 cell clones. We therefore examined more than 70 individual BHK21 cell clones for differences in methylation patterns in the investigated IAPI segment both by HpaII and HhaI restriction patterns and by bisulfite genomic sequencing. Differences in patterns were not detectable (Remus et al. 1999). Of course, we could not scrutinize thousands of individual cell clones for homogeneous methylation patterns. Nevertheless, on the basis of the available evidence, the preferred interpretation of a causative effect of foreign DNA integration on methylation patterns in *trans* will be pursued in future experimental projects.

We also entertained the possibility that the abortive infection of BHK21 cells with Ad12 (Doerfler 1969; Schiedner et al. 1994) with the transcription and expression of exclusively early Ad12 gene products (Hösel et al. 2003) might have affected the stability of cellular DNA methylation patterns. At least on a time scale of days after Ad12 infection, changes in patterns of IAPI genome methylation could not be documented in BHK21 cells (Heller et al. 1999). Productively Ad12-infected human or Ad2-infected hamster cells will also be examined for global changes in methylation patterns in the cellular genomes.

The method of methylation-sensitive representational difference analysis (MS-RDA) is based upon a subtractive hybridization protocol after selecting against DNA fragments that were heavily methylated and, therefore, not cleaved by the HpaII restriction endonuclease (Ushijima et al. 1997). We applied this method to the investigation of transcripts from bacteriophage ë DNA-transgenic hamster cell lines in comparison to hamster cell lines devoid of integrated ë DNA (Müller et al. 2001). By using the suppressive hybridization technique for the analysis of complementary (c)DNA preparations from non-transgenic, Ad12 DNA-transgenic and ë DNA-transgenic hamster cells, several cellular genes were cloned that had altered transcriptional profiles in the transgenic as compared to the non-transgenic cells. Among individual non-transgenic hamster cell clones investigated as negative controls, no differences in cDNA isolates, and hence transcriptional profiles, were observed. We a lso studied these changes in one ë DNA-transgenic mouse strain: Hypermethylation was found for the imprinted *Igf2r* gene for DNA from heart muscle. Two mouse lines transgenic for an Ad2 promoter-indicator gene construct showed hypomethylation in the *interleukin 10* and *Igf2r* genes. We concluded that in Ad12 DNA- or ë DNA-transgenic hamster cells or mice, cellular methylation and transcription patterns can be critically altered (Müller et al. 2001).

Detailed investigations on the heterogeneity of DNA transcription patterns in about 1,170 genes among individual clones of BHK21 and T637 cells have revealed only minimal differences in 5 of these genes by DNA array analyses between the two cell lines and among different clones of each cell line (N. Hochstein and W. D o e r fler, unpublished experiments).

Since the insertion of foreign DNA into established mammalian genomes has become a preferred regimen in experimental biology, e.g., in the generation of transgenic organisms, and increasingly also in gene therapy, I consider it an important problem to pursue the unanticipated,likely unwanted,effect of foreign DNA integration on the stability of the recipient genome. Alterations of patterns of DNA methylation might be merely one—but an experimentally recognizable—manifestation of this disturbance (Doerfler et al. 2001). These problems will be of considerable relevance for certain regimens in gene therapy using the fixation of foreign DNA in an established human genome. When retroviral gene transfer vectors were used to chromosomally fix the human adenine deaminase gene in children with hereditary immunodeficiency, rare T cell leukemias developed (Hacein-Bey-Abina et al. 2003). In these cases, I consider the insertion of foreign DNA, with its consequences for cellular methylation and transcription patterns, as one of the decisive factors explaining this unfortunate outcome of a well-intended medical procedure.

## Towards a Working Hypothesis on Viral Oncogenesis

Viral oncogenesis is frequently accompanied by the integration of the viral genome into the genome of the transformed cell. Integration of viral DNA is a *conditio sine qua non* for transformation (Doerfler 1968,1970) in ells transformed by adenoviruses, by SV40, polyoma virus, by the papilloma viruses HPV 16 and 18, and by retroviruses.

Integration is an important mode of chromosomal fixation and continued expression of the viral genome in the transformed cell. In retroviral replication, proviral integration is an essential step in the viral life cycle. Conventionally, major attention has been directed towards the function of the expressed viral gene products to explain the mechanism of viral oncogenesis.

Having identified the viral "culprit" does not exclude the possibility that the real action is somewhere else, namely in its direct effect on the recipient genome. For some time, we have pursued the possibility that the alterations of DNA methylation patterns enacted in the wake of viral DNA insertion are a general phenomenon following the insertion of any foreign DNA (Doerfler 1995,1996, 2000, Doerfler et al. 2001). Altered cellular methylation patterns then might be a good indicator of more general perturbations in the cellular genome that reach far beyond the immediate site of viral DNA integration. Furthermore, altered methylation patterns forebode changes in transcriptional patterns as well.

Hence, upon foreign DNA insertion, the recipient genome has undergone dramatic functional alterations that might well be at the center of the oncogenic transformation process. Using the Ad12 hamster tumor system as a very efficient experimental model, we have only begun to document changes in cellular transcription patterns in Ad12-induced tumors (Hohlweg et al. 2003).

## Studies on Transgenic Mice: Stability of Patterns of DNA Methylation and Genetic Background in Different Strains of Mice

A construct consisting of the E2A late promoter of Ad2 DNA and the CAT indicator gene was integrated in the non-methylated or in the 5'-CCGG-3' premethylated form into the genome of mice, and the state of methylation was analyzed by HpaII cleavage of DNA from various organs of the transgenic animals (Lettmann et al. 1991). In general, the transgenic construct remained stably integrated. In the founder animal, the non-methylated construct became de novo methylated at all or at most of the 5'-CCGG-3' sites. Pre-imposed methylation patterns were stable for up to four generations beyond the founder animal. However, in the DNA from testes of two founder animals and two F1-males, the premethylated transgenic DNA was demethy-lated by an unknown mechanism. In all other organs, the transgenic DNA preserved the pre-imposed 5'-CCGG-3' methylation patterns. Differences in these transmission modes were not seen depending on whether the transgene was inherited maternally or paternally (Lettmann et al. 1991).

There are studies to support thenotion that genetic background in mice can have a decisive influence on the type of de novo methylation patterns imposed on a foreign DNA transgene and on their stability (for overviews, see Sapienza et al. 1989; Reik et al. 1990; Engler et al. 1991). The molecular mechanisms involved in the "modifier gene" effects are not understood.

We addressed this problem by introducing into the genomes of different mouse strains—DBA/2, 129/sv FVB/N or C57BL/6, CB20, or Balb/c—a construct that consisted of the E2A late promoter of Ad2 DNA and the chloramphenicol acetyltransferase (CAT) gene as reporter. The patterns of de novo transgene methylation were transmitted to the offspring and remained stable for 11 backcross generations, regardless of the heterozygosity in the recipient mouse strain and the presence of presumptive modifier genes. In 7 additional mouse strains carrying the same transgene in different chromosomal locations, strain-specific alterations of methylation patterns were not observed (Schumacher et al. 2000).

We also investigated the stability of DNA methylation patterns in the *Snurf/Snrpn* imprinted gene cluster in mouse embryonal stem cell lines cultured under different experimental conditions, like prolonged passaging, trypsinization, mechanical handling, single cell cloning, staurosporine-induced neurogenesis (Schumacher et al. 2003), or the insertion of foreign (viral) DNA into the ES cell genome. None of these in vitro manipulations affected the stability of the methylation patterns in the analyzed gene cluster (Schumacher and Doerfler 2004). Growth-related genes, *Igf2, H19, Igf2r,* and *GrblO,* are known to respond by altered imprint patterns. The analyzed neuronal gene cluster, however, exhibited stable patterns of DNA methylation under the experimental conditions chosen.

## Fate of Food-Ingested Foreign DNA in the Gastrointestinal Tract of Mice

The tempting interpretation that DNA methylation, particularly the de novo methylation of integrated foreign DNA, is part of an ancient cellular defense mechanism raises a number of questions. One of the obvious ones relates to the major origins of foreign DNA, e.g., in mammals. Virus infection as such a contingency has been extensively discussed in this chapter. Another apparent source of large amounts of foreign DNA that all organisms are constantly exposed to is the DNA orally ingested with the food supply. We have therefore undertaken a study on the fate of food-ingested foreign DNA in mice as model organism. I will present a short summary of the major results my laboratory produced in a project that we initiated in 1988.

In mammals, the gastrointestinal tract is the main portal of entry for foreign macromolecules, and its epithelial lining presents the immediate sites of contact with foreign DNA and proteins. In our investigations on the fate of foreign DNA in the digestive tract, we fed naked test DNA of various derivations to laboratory mice at between 2 and 6 months of age (Schubbert et al. 1994, 1997, 1998). The DNA of bacteriophage M13, the DNA of human Ad2, or the gene for the green fluorescent protein (GFP) from *Aequorea victoria* were administered as test DNA in different experiments. None of these DNA had homologies to bacterial or mouse DNA, except for perhaps very short stretches of DNA sequence that were then excluded from being used for the detection of the foreign DNA in the mouse organism. In later experiments, we fed leaves of the soybean plant to mice and followed the fate of the strictly plant-specific Rubisco (ribulose 1,5-biphosphate carboxylase) gene.

During the passage through the gastrointestinal tract of mice, the bulk of the administered DNA is completelydegraded. However, a small percentage of the test DNA resists the digestive regimens of the gut and can be recovered for several hours after feeding in various parts of the intestinal tube as fragments between 1,700 nucleotides (nt) (rare) and a few 100 nt. By applying a variety of techniques—Southern blotting, PCR, FISH, and rescue of the test DNA fragments by recloning and resequencing—the test DNA could be followed to the wall of the intestinal tract, particularly the colon, Peyer's patches, peripheral white blood cells, and cells of the liver and spleen (Schubbert et al. 1994, 1997, 1998; Hohlweg and Doerfler 2001).

When pregnant animals were test DNA fed, fragments of the test DNA could be traced by FISH and PCR to clusters of cells in various organs of the embryo, but never to all its cells. Moreover, when mice were fed daily and continuously for 8 generations, transgenic animals were never observed. Hence, we assume that the germline must be protected from the exposure to and the uptake of food-ingested foreign DNA. Moreover, we never obtained evidence for the test DNA being transcribed in any of the organ systems of the adult animals that had been given test DNA (Hohlweg and Doerfler 2001). The possible transcription of test DNA was assessed by RT-PCR, the most sensitive technique to detect trace amounts of specific transcripts.

Afterfeedingmicedailyfor 1week, test DNA couldbe recloned—extremely rarely, however—from the spleen of the animals. In a few of these clones, mouse specific DNA was found adjacent to the test DNA in the recloned DNA.

Further proof will be required to investigate the possibility of whether foreign DNA could be integrated into the genome of defense cells in the recipient animals (Schubbert et al. 1997).

In a completely independent approach, we could demonstrate that the protein glutathione-*S*-transferase, a rather stable protein, survived in the stomach and small intestine of mice for up to 30 min after feeding (Palka-Santini et al. 2003).

Ta ken together, the results of this series of investigations indicate that foreign macromolecules, particularly the very stable DNA, can survive in the gastrointestinal tract at least transiently in small amounts and in fragmented form and can get access to various organ systems of the mouse. Even stable proteins survive albeit only for a very short time in the gastrointestinal tube. We have not found any evidence for the entry of foreign DNA into the germ line, nor could we demonstrate transcription of foreign DNA in any of the organ systems tested. It is not known whether a tiny proportion of the thus persisting DNA may find entry into the genome of a rare defense cell and remain there with unknown functional consequences. These questions will be worth pursuing.

## Synopsis and Conclusions

It appears that the following data and interpretations presented in this review have stood the test of time. Research, of course, is a never-ending enterprise, and conclusions have always to be considered subject to change as new data and concepts are being adduced. Here is a synopsis of concepts on the biological significance of 5-mC in the genome the author feels reasonably certain about at the time of this writing (March/November 2005).

1. The virus particle (virion)-encapsidated genomes of most mammalian DNA viruses are not methylated. Likewise, cellular DNA haphazardly integrated into an adenoviral genome, which becomes virion enclosed and does not become methylated, irrespective of its methylation status in the genome it has been derived from. In contrast, the DNA of FV3, an iridovirus, is extensively, p robably completely, methylated.
2. The concept of sequence-specific promoter methylations being causally related to long-term gene silencing, first deduced from work on adenovi-ral promoters, has proved to be generally applicable in most eukaryotic genomes. The promoters of FV3 and of some erythrocyte membrane genes are an interesting exception to this apparent rule.
3. Concomitantly with promoter methylation, histone modifications and perhaps modifications of additional proteins involved in chromatin structure play a decisive role in the regulation of promoter activity. At this time, it seems undecided whether DNA or protein modifications initially orchestrate these regulatory processes. It is likely that a refined interplay between both biochemical mechanisms comes close to the correct answer.
4. Foreign DNA, which has become integrated into an established mammalian genome, often becomes de novo methylated in distinct patterns. The sites of initiation of de novo methylation at least in integrated Ad12 genomes are located paracentrally

in the transgenomes and not close to the junctions with cellular DNA. In integrated Ad12 genomes, this localization of methylation initiation sites might be influenced by the transcriptional activity of the terminally located E1 and E4 regions of the Ad12 genome in the transformed cell lines or in Ad12-induced tumors that are selected for the genetic activity of these viral genome segments. In any event, subsequent to initiation, de novo methylation extends continuously across the transgenomes in a spreading reaction. Initiation seems regional and does not emanate from a specific 5'-CG-3' dinucleotide.

5. It is likely that hypermethylated or rather completely methylated transgenomes are more stably integrated than less completely methylated foreign DNA molecules.
6. At the immediate sites of foreign DNA insertion, the patterns of cellular DNA methylation can be altered extensively.
7. Alterations of cellular DNA methylation patterns are, however, not restricted to the cellular junction sites (Lichtenberg et al. 1987, 1988) but involve remote areas of the recipient genomes, even loci on different chromosomes. This *trans* effect is most striking in retrotransposon sequences, like the endogenous IAP DNA sequences in hamster cells, but can affect genuine cellular sequences as well. These remote perturbations of methylation patterns are not only observed after the integration of Ad12 DNA, which is partly transcriptionally active, but also after insertion of transcriptionally inactive bacteriophage *X* genomes. Possibly, ancient retrotransposons might be more responsive to local alterations of chromatin structure due to foreign DNA insertions into the recipient genome.
8. There is evidence that in addition to alterations of methylation patterns in *trans,* the insertion of foreign DNA could also alter transcriptional patterns in the recipient genomes.
9. Many of the notions summarized here hold true not only for mammalian organisms but also for other eukaryotic genomes, particularly for those of plants.

10. In mammalian genomes, distinct patterns in the distribution of 5-mC residuesexistwhich,atleastinhumans,can be interindividuallypreserved in several (many?) genome segments. These patterns are specific for each genome segment and can be different from cell type to cell type. These observations constitute a major challenge for the Human Epigenome Project.
11. The biological importance of these patterns, which have obviously been conserved over long periods of time, has not been clarified. Long-term gene silencing and chromatin structure as well as the defense against foreign retrotransposons may be factors of significance in explaining the nature of these patterns of genome methylation.
12. Food-ingested foreign DNA persists transiently, in tiny amounts, and in fragmented form during the gastrointestinal passage. Spurious amounts of this DNA can be detected, again transiently, in several organ systems. Transcription of this DNA has not been found.

6

# Establishment and Maintenance of DNA Methylation Patterns in Mammals

In mammals, CpG methylation patterns are established and maintained during development by the Dnmt1 and Dnmt3 families of DNA methyltransferases. These enzymes share conserved catalytic motifs in their C-terminal regions, but have unique N-terminal regulatory domains. Studies over the past several years have shed light on the molecular mechanisms by which DNA methylation patterns are regulated. This review focuses on recent advances in defining the functional domains of DNA methyl-transferases and identifying interacting proteins that may contribute to the functional specializations of these enzymes.

## Abbreviations

| | |
|---|---|
| AdoMet | *S*-Adenosyl-L-methionine |
| APL | Acute promyelocytic leukemia |
| BAH | Bromo-adjacent homology |
| ChIP | Chromatin immunoprecipitation |
| Co-IP | Co-immunoprecipitation |
| DMAP1 | Dnmt1-associated protein |
| Dnmt | DNA methyltransferase |
| ES | Embryonic stem |
| HDAC | Histone deacetylase |
| HMT | Histone methyltransferase |

| | |
|---|---|
| HP1 | Heterochromatin protein 1 |
| ICF | Immunodeficiency, centromeric instability, and facial anomalies |
| MBD | Methyl-CpG-binding domain protein |
| MEF | Mouse embryonic fibroblast |
| PCNA | Proliferating cell nuclear antigen |
| PHD | Plant homeodomain |
| PWWP | Proline-tryptophan-tryptophan-proline |
| RFT | Replication foci-targeting |
| TRD | Transcription repressor domain |
| TSA | Trichostatin A |

## INTRODUCTION

In mammalian somatic cells, cytosine methylation occurs in 60%–80% of all CpG dinucleotides in the genome. However, methylated cytosines are not randomly distributed in the genome, but rather are compartmentalized within specific regions. Heterochromatin, including centromeric and pericentric re-gions,transposableelements,andrepetitive sequencesare heavily methylated. In contrast, CpG islands, which are GC-rich regions that contain high densities of CpG dinucleotides and are located at the promoter regions of many genes, are usually methylation-free. Obvious exceptions to this rule include CpG islands associated with the inactive X chromosome and the silenced al-lele of imprinted genes, which are generally methylated, and this methylation is essential for maintaining the silenced state (Beard et al. 1995; Li et al. 1993; Panning and Jaenisch 1996).

In mice (and probably other mammals as well), DNA methylation patterns are established during embryonic development through a highly orchestrated process that involves genome-wide as well as gene-specific demethylation and de novo methylation. During preimplantation development, both the paternal and the maternal genomes undergo a wave of demethylation, which erases the methylation marks (except those at imprinted loci) inherited from the gametes. After implantation, the embryo undergoes a wave of de novo methylation that establishes a new embryonic methylation pattern, which is maintainedin somatic tissues (Jaenisch 1997; JaenischandBird 2003; Li 2002). Demethylation and de novo methylation also occur during gametogenesis and have been shown to playa criticalrole in the establishment of parental-specific methylation marks in imprinted loci (Reik et al. 2001).

Four DNA methyltransferases, namely Dnmt1, Dnmt2, Dnmt3a, and Dnmt3b, have been identified in humans and mice (Bestor et al. 1988; Okano et al. 1998a, b; Van den Wyngaert et al. 1998; Xie et al. 1999; Yoder and Bestor 1998). Genetic manipulation of these enzymes in mice has demonstrated that DNA methylation is essential for mammalian development and plays crucial roles in a variety of biological processes, such as gene regulation, genomic imprinting, and X chromosome inactivation. Studies over the last several years have focused on understanding the molecular mechanisms by which DNA methylation patterns are regulated. In this chapter, we discuss the structural features and interacting proteins of DNA methyltransferases that may contribute to the functional specialization of these enzymes.

## DNA METHYLTRANSFERASES

### Dnmt1

Dnmt1 is expressed constitutively in proliferating cells and ubiquitously in somatic tissues throughout mammalian development. Purified Dnmt1 protein methylates DNA containing hemimethylated CpG dinucleotides more efficiently than unmethylated DNA in vitro (Pradhan et al. 1999; Yoder et al. 1997).

The Dnmt1 protein has been shown to localize to DNA replication foci during S phase (Leonhardt et al. 1992), indicating that its function is coupled to DNA replication. Inactivation of the mouse *Dnmt1* gene by gene targeting results in extensive demethylation of all sequences examined, but has little effect on de novo methylation of newly integrated retrovirus DNA (Lei et al. 1996; Li et al. 1992). Furthermore, overexpression of Dnmt1 alone fails to induce de novo methylation in mouse embryonic stem (ES) cells or in *Drosophila* (Chen et al. 2003; Lyko et al. 1999). These findings suggest that Dnmt1 functions primarily as a maintenance methyltransferase, which copies the parental-strand methylation pattern onto the daughter strand after each round of DNA replication, and that it alone has little or no de novo methyltransferase activity.

## Dnmt2

Dnmt2 is a member of a protein family conserved from yeast (*Schizosaccha-romyces pombe*) and fruit fly (*Drosophila melanogaster*) to mammals. The yeast pmt1 has been demonstrated to be enzymatically inactive due to an amino acid change at the catalytic site (Pinarbasi et al. 1996; Wilkinson et al. 1995), whereas the Dnmt2 homologs from *Drosophila* (dDnmt2), mouse (mDnmt2), and human (hDnmt2) all contain the conserved DNA methyl-transferase motifs (Hung et al. 1999; Lyko et al. 2000; Okano et al. 1998b; Tweedie et al. 1999; Van den Wyngaert et al. 1998; Yoder and Bestor 1998). hDnmt2 has been shown to form a two-domain structure that is similar to that of M.HhaI, an active prokaryotic cytosine methyltransferase (Dong et al. 2001).

Recent studies have shown that dDnmt2, mDnmt2, and hDnmt2 are all genuine DNA methyltransferases, and their primary targets appear to be non-CpG sites (Hermann et al. 2003; Kunert et al. 2003; Liu et al. 2003; Tang et al. 2003). While the biological function of Dnmt2 remains to be determined, genetic studies have demonstrated that it is not essential for *Drosophila* and mammalian development (Kunert et al. 2003; M. Okano andE. Li, unpublished results).

## Dnmt3

The Dnmt3 family has three members: Dnmt3a, Dnmt3b, and Dnmt3L (Dnmt3-like). Dnmt3L, which shows sequence similarity to Dnmt3a and Dnmt3b, does not have enzymatic activity due to the lack of some critical catalytic motifs, but may function as a regulator of DNA methylation (see Sect. 3.2.11; Aapola et al. 2001; Hata et al. 2002). Unlike Dnmt1, Dnmt3a and Dnmt3b are highly expressed in ES cells, early embryos, and developing germ cells, where de novo methylation is known to take place, but are downregulated in somatic tissues of postnatal animals (Okano et al. 1998a). Recombinant Dnmt3a and Dnmt3b proteins methylate both unmethylated and hemimethylated DNA substrates with similar efficiency in vitro (Aoki et al. 2001; Okano et al. 1998a). Inactivation of both *Dnmt3a* and *Dnmt3b* by gene targeting blocks de novo methylation in ES cells and early embryos (Okano et al. 1999).

Recently, Dnmt3a and Dnmt3L, but not Dnmt3b, have been shown to be essential for the establishment of methylation imprints during gametogenesis (Bourc'his et al. 2001; Hata et al. 2002; Kaneda et al. 2004). Moreover, Dnmt3a and Dnmt3b cause de novo methylation when overexpressed in mammalian cells or transgenic flies (Chen et al. 2003; Hsieh 1999; Lyko et al. 1999). These findings strongly support the notion that Dnmt3a and Dnmt3b function primarily as de novo methyltransferases, which are responsible for the establishment of DNA methylation patterns during embryogenesis and gametogenesis.

In addition to establishing DNA methylation patterns, Dnmt3a and Dnmt3b play a role in maintaining global DNA methylation levels as well. Based on genetic studies in mice, the Dnmt1 and Dnmt3 families of methyltransferases have distinct and non-redundant functions but they act cooperatively to maintain hypermethylation of the genome. While Dnmt1 is the major maintenance enzyme, the relative contributions of Dnmt1, Dnmt3a, and Dnmt3b to the maintenance of global methylation appear to differ in a cell type-specific manner. Targeted disruption of *Dnmt3a* or *Dnmt3b* in mouse ES cells has only minor effect on global methylation (despite demethylation of specific sequences), whereas disruption of both *Dnmt3a* and *Dnmt3b* results in progressive loss of methylation throughout the genome, indicating that the two enzymes have largely redundant functions in ES cells (Chen et al. 2003; Liang et al. 2002).

In mouse embryonic fibroblasts (MEFs), however, Dnmt3b, but not Dnmt3a, is required for the maintenance of DNA methylation levels (Dodge et al. 2005). Unlike *Dnmt3a*$^{-/-}$ *Dnmt3b*$^{-/-}$ ES cells, *Dnmt3a*$^{-/-}$ *Dnmt3b*$^{-/-}$ MEFs do not show progressive loss of methylation in culture, suggesting that in the absence of Dnmt3a and Dnmt3b, Dnmt1 is capable of maintaining a higher level of DNA methylation in MEFs than in ES cells (Dodge et al. 2005). Genetic studies of human DNMTs have also been carried out in various cancer cell lines. It has been reported that the HCT116 colon cancer cells lacking DNMT1 or DNMT3B retain significant genomic methylation and associated gene silencing, whereas cells with both DNMT1 and DNMT3B inactivated show much lower levels of DNA methylation, suggesting that the two enzymes function redundantly to maintain CpG methylation (Rhee et al. 2000, 2002).

However, recent studies have shown that depletion of DNMT1 alone by either antisense or siRNA in HCT116 cells and other human cancer cells results in global and gene-specific demethylation and re-expression of tumor suppressor genes (Robert et al. 2003). Further studies are necessary to determine whether normal and cancer cells use different mechanisms to maintain DNA methylation patterns.

## REGULATION OF DNA METHYLATION PATTERNS

One of the challenges in the field is to understand the molecular mechanisms by which DNA methylation patterns are regulated. Since the Dnmt1 and Dnmt3 families of methyltransferases do not appear to have any sequence specificity beyond CpG dinucleotides (Dodge et al. 2002; Okano et al. 1998a; Yoder et al. 1997), chromatin-based mechanisms have been proposed to explain how DNA methyltransferases find their target sequences in the genome (Bird 2002).

Establishment and maintenance of DNA methylation patterns is most likely accomplished by molecular interactions involving the DNA methyltransferases and other chromatin-associated factors. Indeed, a number of proteins have been identified that interact with one or more Dnmts, and many of these proteins are involved in the regulation of chromatin structure and gene expression (Table 1).

### Structural Features of DNA Methyltransferases

The catalytic domain of cytosine methyltransferases contains 10 characteristic sequence motifs, 6 of which are highly conserved. X-ray crystallography studies indicate that the catalytic domain is organized into a two-domain structure.

The large domain encompasses most of the conserved motifs and contains the catalytic center and the binding pocket for the co-factor *S*-adenosyl-l-methionine (AdoMet). The small domain comprises the variable region and may be partially responsible for DNA target recognition (Cheng and Roberts 2001; Dong et al. 2001).

In addition to the catalytic domain, the Dnmt1 and Dnmt3 proteins have unique N-terminal extensions, which are believed to be the structural basis for the differences in biochemical properties and biological functions exhibited by these enzymes. As the major maintenance DNA methyltransferase, Dnmt1 methylates hemimethylated CpGs and its action is coupled to DNA replication. By mutagenesis analysis, a region within the N terminus of Dnmt1 was initially identified as the replication foci-targeting (RFT) domain (Leonhardt et al. 1992). Later studies revealed that the proliferating cell nuclear antigen (PCNA)-interacting domain and the bromo-adjacent homology (BAH) domain also contribute to the association of Dnmt1 with the DNA replication machinery during the S phase (Chuang et al. 1997; Liu et al. 1998).

The location of the target recognition domain, the domain responsible for recognizing hemimethylated CpG sites, is still controversial. Fatemi et al. showed that the catalytic domain has a preference for binding hemimethylated CpG sites (Fatemi et al. 2001), whereas Araujo et al. mapped the target recognition domain to a region in theNterminus (aminoacids122–417,inproximityto the PCNA-interacting domain) (Araujo et al.2001). TheCXXC domain, a cysteine-rich $Zn^{2+}$-binding motif located between the RFT and BAH domains, has also been shown to bind DNA sequences containing CpG dinucleotides and, thus, may be involved in target recognition as well (Chen and Li 2004).

## DNMT-LNTERACTING PROTEINS

### HDACs

HDACs are enzymes that catalyze the removal of acetyl groups from lysine residues in both histone and non-histone proteins. They play a key role in the regulation of gene transcription and many other biological processes involving chromatin. It has been well established that DNA methylation represses transcription partly by recruitment of co-repressor complexes containing HDACs via methyl-CpG-binding proteins (MBDs) (see the following section; Bird and Wolffe 1999). More recent evidence, however, indicates additional connections between DNA methylation and histone deacetylation. A number of studies have demonstrated that the Dnmt1 and Dnmt3 families of methyltransferases directly interact with one or more HDACs. Dnmt1 has been shown to bind HDAC1 and HDAC2 via its N-terminal regulatory region, and Dnmt3a, Dnmt3b,and Dnmt3L have been shown to bind HDAC1 through their ATRX-homology domain.

These findings suggest a possible feedback loop between DNA methylation and histone acetylation whereby each modification reinforces the other, thus creating a stably silenced chromatin state. Consistent with this notion, the demethylating agent 5-aza-2'deoxycytidine (5-aza-dC) and the HDAC inhibitor trichostatin A (TSA) are synergistic in reactivating gene expression in cancer cells (Cameron et al. 1999). Although the N-terminal regulatory domains of the Dnmt1 and Dnmt3 proteins have been shown to function as transcriptional repressors in reporter assays (Aapola et al. 2002; Bachman et al. 2001; Deplus et al. 2002; Fuks et al. 2000, 2001; Robertson et al. 2000; Roun-tree et al. 2000), currently there is no convincing evidence that the Dnmts repress transcription independent of their DNA methyltransferase activities.

## MBDs

A family of five proteins (MeCP2, MBD1, MBD2, MBD3, and MBD4), has been identified as binding proteins to methyl-CpG sequences (Hendrich and Bird 1998; Lewis et al. 1992; Meehan et al. 1989). These proteins are characterized by having a common structural domain of roughly 70 amino acids named the methyl-CpG-binding domain (MBD). With the exception of MBD4, all MBD family members are involved in transcriptional repression, presumably by recruiting chromatin-remodeling complexes to methylated CpG sites (Bird and Wo lffe 1999). For instance, MeCP2 has been shown to interact with the Sin3a-HDAC co-repressor complex (Jones et al. 1998; Nan et al. 1998), and MBD2 and MBD3 have been shown to associate with the NuRD (nucle-osome remodeling and histone deacetylation) complex, which contains an ATP-dependent chromatin-remodeling protein, Mi-2, and HDACs (Feng and Zhang 2001; Wade et al. 1999).

In addition to acting downstream of DNA methylation to "interpret" the methylation signal, some MBDs have been shown to directly interact with Dnmt1. Tatematsu et al. showed that the MBD2–MBD3 complex binds to both hemimethylated and fully methylated

DNA in vitro, co-localizes with DNMT1 at replication foci in 293 cell nuclei at late S phase, and associates with DNMT1 by co-immunoprecipitation (co-IP) assay (Tatematsu et al. 2000). Subsequently, Kimura and Shiota showed that MeCP2 also interacts with Dnmt1 and forms complexes with hemimethylated as well as fully methylated DNA (Kimura and Shiota 2003). The region of MeCP2 that interacts with Dnmt1 corresponds to the transcription repressor domain (TRD), which also recruits HDACs via Sin3a. Interestingly, the MeCP2-Dnmt1 complex does not contain HDAC1, suggesting that Dnmt1 and the Sin3a-HDAC1 complex may compete for MeCP2 (Kimura and Shiota 2003).These observations led to the hypothesis that MBDs may playa role in targeting Dnmt1 to hemimethylated DNA during S phase (Kimura and Shiota 2003; Tatematsu et al. 2000). It should be noted, however, that the interactions between Dnmt1 and MBDs were observed in cells overexpressing these proteins or in in vitro assays. Genetic evidence for the involvement of MBDs in maintenance of DNA methylation is still lacking.

## DMAP1

DMAP1 (Dnmt1-associated protein) was identified in a yeast two-hybrid screen using the Dnmt1 N-terminal region as bait. Human DMAP1 consists of 467 amino acids and contains a putative nuclear localization signal (NLS) and a predicted coiled-coil domain. Mutagenesis analysis indicates that the extreme N terminus (126 amino acids) of Dnmt1 interacts with the coiled-coil domain of DMAP1 (Fig. 1; Rountree et al. 2000). Notably, the oocyte-specific form Dnmt1o, which lacks the DMAP1-interacting domain, has equivalent maintenance methylation function as Dnmt1 (Ding and Chail-let 2002), indicating that the Dnmt1-DMAP1 interaction does not affect the methyltransferase activity of Dnmt1. However, Dnmt1o is significantly more stable than Dnmt1 in vivo (Ding and Chaillet 2002), raising the possibility that DMAP1 may be involved in Dnmt1 degradation. In addition, DMAP1 may play a role in targeting Dnmt1 to specific genomic loci by virtue of its ability to interact with various proteins.

Recent studies have shown that DMAP1 mediates the interactions between Dnmt1 and a number of proteins, including TSG101 (a transcriptional co-repressor involved in the silencing of nuclear hormone-induced genes), Daxx (a protein involved in a variety of processes, including apoptosis and transcriptional regulation), and RGS6 (an RGS family member involved in the regulation of heterotrimeric G protein signaling) (Liu and Fisher 2004; Muromoto et al. 2004; Rountree et al. 2000).

## Rb

The retinoblastoma protein (Rb), a member of the "pocket proteins" family, is a major cell cycle regulator, and its inactivation is associated with the development of retinoblastoma and other types of human cancers. Rb exerts its anti-proliferative activity primarily by inhibiting the E2F transcription factors (Weinberg 1995). Binding of Rb to E2F is thought to mask the transactivation domain of E2F. In addition, Rb has been shown to actively repress transcription at E2F-responsive promoters by recruiting chromatin remodeling factors such as HDACs, HMTs, and members of the ATP-dependent chromatin remodeling complex SWI/SNF (Brehm et al. 1998; Dunaief et al. 1994; Luo et al. 1998; Magnaghi-Jaulin et al. 1998; Nielsen et al. 2001; Trouche et al. 1997).

By biochemical fractionation, Robertson et al. showed that DNMT1 is also a component of the Rb-E2F1 complex (Robertson et al. 2000). The Rb-DNMT1 interaction was later confirmed by another study (Pradhan and Kim 2002).

The interaction domains were mapped to the N-terminal region of DNMT1 and the A, B, and C pockets of Rb (Fig. 1; Pradhan and Kim 2002; Robertson et al. 2000). Reporter assays showed that DNMT1 cooperates with Rb to repress transcription from promoters containing E2F-binding sites. However, the reporter construct re-isolated from transfected cells was not methylated, suggesting that DNA methylation may not be involved in this repression (Robertson et al. 2000). Indeed, a separate study showed that Rb inhibits the methyltransferase activity of DNMT1 by interfering with the formation of DNMT1-DNA binary complex (Pradhan and Kim 2002). It would be interesting to determine whether Rb prevents some CpG sites from being methylated in normal cells and whether its loss contributes to aberrant methylation patterns in cancer cells.

## PCNA

PCNA, a ring-shaped protein known as a processivity factor for DNA poly-merase ä, is involved in DNA replication and repair. In addition to this role, PCNA interacts with a number of other proteins to increase their local concentration at replicated DNA sites. Dnmt1 has a typical PCNA-binding motif in its N-terminal region and has been shown to interact with PCNA.

Immunofluorescence analysis shows precise colocalization of Dnmt1 with early replication foci, and a point mutation within the PCNA-binding motif alters Dnmt1 localization, suggesting that PCNA plays a role in recruiting Dnmt1 to methylate newly replicated DNA (Chuang et al. 1997). In agreement with this notion, a recent study showed that PCNA facilitates the action of Dnmt1 on hemimethylated DNA in vitro (Iida et al. 2002). The Dnmt1–PCNA interaction is regulated by the cell cycle regulator p21, which binds tightly to PCNA via a sequence similar to the PCNA-binding motif of Dnmt1. A peptide derived from p21 that contains the sequence has been shown to disrupt the Dnmt1-PCNA interaction (Chuang et al. 1997). This observation, along with the recent finding that p53 represses Dnmt1 transcription (Peterson et al. 2003), suggests that the p53 pathway is involved in the regulation of DNA methylation.

## PML-RAR

The promyelocytic leukemia-retinoic acid receptor (PML-RAR) fusion protein is an oncogenic transcription factor found in acute promyelocytic leukemia (APL). Previous studies have demonstratedthatPML-RARrepresses transcription of target genes through the recruitment of HDAC complexes (Grignani et al. 1998; Lin et al. 1998). Di Croce et al. have shown that ectopic expression of PML-RAR in U937 hemopoietic precursor cells leads to promoter hypermethylation and silencing of RARp, a PML-RAR target gene. Consistent with this, blasts from 7 of 9 APL patients and the APL-derived NB4 cell line had methylation of the 5' region of RARp. PML-RAR-induced repression of RARp transcription was partially released by 5-aza-dC, indicating that DNA methylation contributes to the repressive effect of PML-RAR. Co-IP, chromatin IP (ChIP), and localization analyses indicated that PML-RAR forms complexes with DNMT1 and DNMT3a, and the DNMT3a-interacting site was mapped to the PML moiety (Di Croce et al. 2002). These results suggest a possible role for PML-RAR in recruiting DNMTs to its target genes. It should be noted, however, that many malignancies show abnormal methylation of RARp without the PML-RAR translocations, in some cases more commonly than APL. Moreover, the frequency of methylation of other PML-RAR target genes *(pl5$^{INK4b}$, CDH1, andp73)* does not appear to differ between APL patients and those with other subtypes of acute myelogenous leukemia (Esteller et al. 2002).

**Myc**

The Myc transcription factor is involved in the regulation of a variety of cellular processes, including growth, proliferation, differentiation, and apoptosis (Pelengaris et al. 2002). Myc heterodimerizes with a small protein called Max to function as a sequence-specific, DNA-binding transcriptional activator. In addition, Myc has been shown to repress the transcription of specific genes (Patel et al. 2004). The mechanisms by which Myc silences gene expression are not well understood, although Myc has been shown to physically bind to and functionally interfere with certain transcriptional activators, such as Miz-1 (Staller et al. 2001; Wu et al. 2003). A recent report has implicated the involvement of DNA methylation in Myc-mediated gene silencing (Brenner et al. 2005).

Glutathione S-transferase (GST) pull-down and co-IP experiments indicate that Myc interacts with Dnmt3a and associates with DNA methyl-transferase activity. In reporter assays, Dnmt3a enhances Myc-mediated inhibition ofthe *p21Cipl* promoter activity. In U2OS cells with Dnmt3a depleted by antisense RNA, expression of *p21Cipl* is increased. ChIP analysis shows that the *p21Cipl* proximal promoter is bound by Dnmt3a in wild-type *(c-myc$^{+,+}$)* rat fibroblasts and the Dnmt3a binding is significantly reduced in *c-myc* knockout *(c-myc$^{-,-}$)* cells. Furthermore, 5-azacytidine treatment results in elevated expression of *p21Cipl* in *c-myc$^{+,+}$* cells, but has no effect on *p21Cipl* level in *c-myc$^{-,-}$* cells (Brenner et al. 2005). These results suggest that Myc, in cooperation with other factors such as Miz-1, may play a role in targeting DNA methylation to specific genomic regions.

## RP58

RP58 is a 58-kDa protein that contains an N-terminal POZ domain, belonging to the POZ/zinc finger family, and four sets of Kruppel-type zinc-finger motifs in its C terminus. It has been shown to associate with condensed chromatin and mediate sequence-specific transcriptional repression (Aoki et al. 1998). In a yeast two-hybrid screen using the Dnmt3a ATRX-homology domain as bait, RP58 was identified as an interacting partner of Dnmt3a. Further analyses confirmed that RP58, via a region (aa 310–427) that contains two of the Kruppel-type zinc fingers, directly contacts the ATRX-homology domain ofDnmt3a. Currently, thebiologicalsignificance ofthe RP58–Dnmt3a interaction is unknown, although Dnmt3a has been shown to enhance RP58-mediated transcriptional repression in reporter assays (Fuks et al. 2001). Given that deletion of the ATRX-homology domain of Dnmt3a does not significantly alter its cellular localization (Chen et al. 2004), it is unlikely that RP58 plays a major role in targeting Dnmt3a to heterochromatin. However, it would be interesting to determine whether Dnmt3a contributes to the association of RP58 with condensed chromatin.

## Suv39h1 and HP1

Studies in several model organisms have provided genetic evidence that methylation of lysine 9 of histone H3 (H3-K9) controls DNA methylation. In *Neurospora crassa*, a mutation in the H3-K9 methyltransferase gene *dim-5* causes global hypomethylation of DNA (Tamaru and Selker 2001). In *Ara-bidopsis thaliana*, the KRYPTONITE HMT is required for maintaining CpNpG methylation (Jackson et al. 2002; Malagnac et al. 2002). In mouse ES cells, targeted disruption of the H3-K9 methyltransferases Suv39h1 and Suv39h2 results in demethylation of the major satellite repeats at pericentric heterochromatin (Lehnertz et al. 2003).

While the mechanisms underlying this dependence of DNA methylation on histone methylation is not well understood, HP1—which binds with high affinity to H3 when methylated at K9 (Bannister et al. 2001; Lachner et al. 2001)—may provide a molecular link. Disruption of the *Neurospora* HP1 homolog results in complete loss of DNA methylation, whereas a mutation in the *Arabidopsis* HP1 homolog does not appear to affect DNA methylation (Freitag et al. 2004; Malagnac et al. 2002). Genetic evidence for the involvement ofHP1 in DNA methylation in mammals is still lacking. However, recent studies indicate that Dnmt3a and Dnmt3b associate with HP1 (Fuks et al. 2003; Lehnertz et al. 2003). Dnmt3a and Dnmt3b have been shown to colocalize with HP1 at heterochromatic foci (Bachman et al. 2001; Chen et al. 2004).

Such a localization pattern seems to be dependent on H3-K9 methylation, as Dnmt3b and HP1 fail to concentrate at heterochromatic foci in Suv39h1 and Suv39h2 double knockout cells (Lehn-ertz et al. 2003). Co-IP experiments show that Dnmt3a and Dnmt3b form complexes with HP1, apparently in a Suv39h-independent manner (Fuks et al. 2003; Lehnertz et al. 2003). Dnmt3a and Dnmt3b have also been shown to associate with H3-K9 methyltransferase activity (Fuks et al. 2003; Lehnertz et al. 2003). One study shows that Dnmt3a, via its ATRX-homology domain, directly interacts with Suv39h1. A separate study shows that the H3-K9 methyltransferase activities associated with Dnmt3b in wild-type and Suv39h double knockout cells are equally robust, suggesting that Dnmt3b forms one or more histone-DNA methylation complexes containing Suv39h-unrelated H3-K9 methyltransferases (Lehnertz et al. 2003).

## SUMO-1, Ubc9, PIAS1, and PIASxá

The small ubiquitin-related protein SUMO-1 posttranslationally modifies many proteins with roles in diverse processes including regulation of transcription, chromatin structure, and DNA repair. SUMO-1 is ligated to lysine residues in substrate proteins via a three-step enzymatic process involving a heterodimeric E1 activating enzyme (SAE1/SAE2), an E2 conjugating enzyme (Ubc9), and a number of E3 ligating enzymes (PIAS proteins, RanBP2, andPc2). In contrast to ubiquitination, sumoylation does not promote protein degradation but instead modulates several other aspects of protein function, including subcellular localization, protein–protein interactions, protein–DNA interactions, and enzymatic activity (Gill 2004).

Using yeast two-hybrid screens, two groups have identified several components of the sumoylation machinery as Dnmt3a- and Dnmt3b-interacting partners. These include Ubc9, PIAS1, and PIASxá. The interactions are further confirmed by co-localization, co-IP, and GST pull-down experiments. Mutagenesis analyses map the interaction domain to the N-terminal regions of Dnmt3a and Dnmt3b. Dnmt3a and Dnmt3b can be sumoylated when co-transfected with SUMO-1 in cells or when incubated with recombi-nant E1 (SAE1/SAE2), Ubc9, and SUMO-1 in the presence of ATP (Kang et al. 2001; Ling et al. 2004). In co-transfection experiments, overexpression of SUMO-1 inhibits Dnmt3a-HDAC interaction and relieves Dnmt3a-mediated transcriptional repression of a reporter gene (Ling et al. 2004). These results suggest that sumoylation may regulate the functions of Dnmt3a and Dnmt3b.

## Dnmt3L

As discussed above, Dnmt3L belongs to the Dnmt3 family, but does not have enzymatic activity. Dnmt3L contains an ATRX-homology domain that is closely related to that of Dnmt3a and Dnmt3b. Its C-terminal region shows sequence homology to the catalytic domain of Dnmt3a and Dnmt3b, but lacks some residues known to be critical for enzymatic activity, including the PC dipeptide at the active site. The expression pattern of *Dnmt3L* is strikingly similar to that of *Dnmt3a* and *Dnmt3b* during mouse development (Hata et al. 2002). Genetic studies have demonstrated that *Dnmt3L*, like *Dnml3a*, is essential for the establishment of genomic imprinting.

Although disruption of *Dnmt3L* in the zygote does not affect embryonic development, *Dnmt3L*$^{-/-}$ females fail to establish maternal methylation imprints in the oocytes, which leads to loss of monoallelic expression of maternally imprinted genes and developmental defects in the offspring, and *Dnmt3L*$^{-/-}$ males show defects in spermatogenesis (Bourc'his and Bestor 2004; Bourc'his et al. 2001; Hata et al. 2002). Dnmt3L has been shown to directly interact with Dnmt3a and Dnmt3b via their C-terminal regions, resulting in stimulation of the catalytic activity of these de novo methyltransferases. In vitro assays show that complex formation between Dnmt3a andDnmt3L accelerates DNA and AdoMet binding to Dnmt3a (Gowher et al. 2005). Moreover, Dnmt3L has been shown to associate with HDAC1 via its ATRX-homology domain and function as a transcriptional re-pressorin reportersystem. Taken together, Dnmt3L may regulate genomic imprinting by enhancing the activity of Dnmt3a or by increasing the accessibility of Dnmt3a to imprinted loci.

## Concluding Remarks

Over the past several years, our understanding of the molecular mechanisms by whichDNA methylation patterns are establishedandmaintained has been growing steadily. The identification of a growing number of chromatin-associatedproteins that interact withoneormoreDnmtssupports thehypoth-esis that chromatin structure and chromatin proteins play important roles in the regulation of the activities and specificities of DNA methyltransferases. It should be noted, however, that many of the Dnmt-interacting partners were identified by candidate approaches or yeast two-hybrid screens. Much needs to be done to verify these interactions. Moreover, with the exception of a few cases such as the Dnmt3a–Dnmt3L interaction, the functional implications of these interactions remain largelyunknown due to the lack of genetic evidence. Another challenge we are facing is how to assemble the individual interacting proteins into regulatory complexes and pathways. In the future, we expect to see more studies that address these issues.

# 7

# Molecular Enzymology of Mammalian DNA Methyltransferases

DNA methylation is an essential modification of DNA in mammals that is involved in gene regulation, development, genome defence and disease. In mammals 3 families of DNA methyltransferases (MTases) comprising (so far) 4 members have been found: Dnmt1, Dnmt2, Dnmt3A and Dnmt3B. In addition, Dnmt3L has been identified as a stimulator of the Dnmt3A and Dnmt3B enzymes. In this review the enzymology of the mammalian DNA MTases is described, starting with a depiction of the catalytic mechanism that involves covalent catalysis and base flipping.

Subsequently, important mechanistic features of the mammalian enzyme are discussed including the specificity of Dnmt1 for hemimethylated target sites, the target sequence specificity of Dnmt3A, Dnmt3B and Dnmt2 and the flanking sequence preferences of Dnmt3A and Dnmt3B. In addition, the processivity of the methylation reaction by Dnmt1, Dnmt3A and Dnmt3B is reviewed. Finally, the control of the catalytic activity of mammalian MTases is describedthat includes the regulation ofthe activity of Dnmt1 by its N-terminal domain and the interaction of Dnmt3A and Dnmt3B with Dnmt3L. The allosteric activation of Dnmt1 for methylation at unmodified sites is described. Wherever possible, correlations between the biochemical properties of the enzymes and their physiological functions in the cell are indicated.

## INTRODUCTION

The first mammalian DNA methyltransferase (MTase) activity was discovered byRazin's group in the early 1980s (Gruenbaum et al. 1982). The enzyme responsible for this activity is called Dnmt1 today [the name derives from *DNA methyltransferase*; the systematic nomenclature of DNA MTases is described in Roberts et al. (2003)]. The murine Dnmt1 enzyme was the first mammalian DNA MTase to be cloned and expressed recombinantly (Bestor et al. 1988; Pradhan et al. 1997). During the last decade, three more members of the mammalian Dnmt enzyme family have been discovered and cloned (Fig. 1; reviews: Chen and Li 2004; Hermann et al. 2004a). All these enzymes contain a domain of approximately 400–500 amino acid residues, which is characterised by the presence of 10 conserved amino acid motifs, shared between prokaryotic and eukaryotic DNA-(cytosine-C5)-MTases (reviews: Cheng 1995; Jeltsch 2002).

The catalytic centre and coenzyme binding site of MTases reside within this domain. In addition, the Dnmt1 and the Dnmt3 enzymes harbour large N-terminal regulatory parts (reviews: Chen and Li 2004; Hermann et al. 2004a). The N-terminal regulatory domain of Dnmt1 contains different motifs and subdomains which interact with many other proteins (Chuang et al. 1997; Fuks et al. 2003; Liu and Fisher 2004; Margot et al. 2003; Pradhan and Kim 2002; Robertson et al. 2000; Rountree et al. 2000). One example of these interacting proteins is the proliferating cell nuclear antigen (PCNA) known as processivity factor for the DNA polymerases å/ä (Chuang et al. 1997; Maga and Hubscher 2003). It seems that the N-terminus is forming a platform for binding of proteins involved in chromatin condensation, gene regulation and DNA replication. In addition, Dnmt1 has a role in mismatch repair of mammalian cells (Kim et al. 2004; Wang and James Shen 2004).

Dnmt1 has a strong preference for methylation of hemimethylated CG sites (Fatemi et al. 2001; Gruenbaum et al. 1982; Hermann et al. 2004b), which implicates it as havinga function in maintenance of the methylation pattern of the DNA after replication. Dnmt1 knock-out mice die during embryogenesis; embryos show almost complete loss of DNA methylation (Li et al. 1992). Interestingly, the catalytic domain of Dnmt1 is inactive in the absence of the N-terminal part (Fatemi et al. 2001), which implies an important regulatory function of the N-terminal domain on the enzyme.

Dnmt2 is the smallest enzyme among the eukaryotic MTases and it comprises only the catalytic domain. It has a very slow turnover rate (Hermann et al. 2003; Kunert et al. 2003; Liu et al. 2003; Tang et al. 2003). The protein is conserved in many eukaryotic species (also some that only have low or even undetectable levels of DNA methylation like *Drosophila melanogaster* or *Schizosaccharomyces pombe*). The biological function of Dnmt2 is not known, although it has been associated to longevity in *D. melanogaster* (Lin et al. 2004).

The mammalian Dnmt3 enzyme family consists of three different proteins, Dnmt3A, Dnmt3B and Dnmt3L. The regulatory N-terminal domain of Dnmt3A and Dnmt3B is not essential for catalysis (Gowher and Jeltsch 2002; Reitheretal. 2003). Both enzymes contain an ATRX-likeCys-rich domain(also called PHD domain) and a PWWP domain, which are involved in interactions with other proteins and targeting to heterochromatin (Aapola et al. 2002; Bachman et al. 2001; Chen andLi 2004; Fuks et al. 2003; Ge et al. 2004). Despite significant amino acid sequence and biochemical similarities, Dnmt3A and Dnmt3B have distinct biological roles. Dnmt3B is responsible for methylation of pericentromeric satellite regions (Hansen et al. 1999;Okano et al. 1999;Xu et al. 1999). Dnmt3B$^{-/-}$ knock-out mice die during the late embryonic stage and the embryos lack methylation in pericentromeric repeat regions (Okano et al. 1999). Loss of Dnmt3B activity in human leads to ICF (immunodeficiency, centromere instability, facial anomalies) syndrome, a genetic disorder that is accompanied by low methylation in the pericentromeric satellite regions of chromosomes 1, 9 and 16 (Ehrlich 2003).

Dnmt3A knock-out mice show developmental abnormalities and die a few weeks after birth (Okano et al. 1999). This enzyme has been associated with the methylation of single copy genes and retrotransposons (Bourc'his and Bestor 2004; Bourc'his et al. 2001; Hata et al. 2002) andit is required for the establishment of the genomic imprint during germ cell development (Kaneda et al. 2004). The N-terminal part of Dnmt3L is shorter than those of Dnmt3A and Dnmt3B and only contains the PHD domain. The C-terminal part of this protein is truncated and all its "catalytic" motifs are crippled, indicating it cannot be an active DNA MTase. Dnmt3L acts as a stimulator of the catalytic activity of Dnmt3A and Dnmt3B activity (Chedin et al. 2002; Gowher et al. 2005; Suetake et al. 2004).

In the following sections, the enzymology of the mammalian DNA MTases will be reviewed. Starting with a description of the catalytic mechanism, some important mechanistic features like thedegree of specificity for the target base and preference for flanking sequences, the processivity of DNA methylation andthe mechanism of control of enzyme activitywill be discussed. Itis written under the presumption that a detailed knowledge of the enzymes' properties is an essential prerequisite for the understanding of their cellular roles.

## CATALYTIC MECHANISM OF DNA-(CYTOSINE-C5)-MTASES

All DNA MTases use the coenzyme *S*-adenosyl-l-methionine (AdoMet) as the source for the methyl group being transferred to the DNA bases. The methyl group of AdoMet is bound to a sulphonium centre, which activates it towards nucleophilic attack. The AdoMet binding site is remarkably conserved in all DNA (and also non-DNA) MTases. It is created by residues from the motifs I–III and X, which form conserved contacts to almost every hydrogen bond donor and acceptor of the AdoMet and, in addition, several hydrophobic interactions to the cofactor. The roles of many of these residues have been confirmedby mutagenesis experiments in prokaryotic MTases (review: Jeltsch 2002).

### Reaction Mechanism of DNA-(Cytosine-C5)-MTases

The reaction mechanism of cytosine-C5 methylation was uncovered for the prokaryotic DNA-(cytosine-C5)-MTase M.HhaI (Fig. 2; Wu and Santi 1985; Wu a n d Santi 1987). A key feature of the catalytic process is a nucleophilic attack of the enzyme on the carbon-6 of the target cytosine. This attack is performed by the thiol group of the cysteine residue that is part of the conserved PCQ motif in the active site of cytosine-C5-MTases (motif IV). This reaction is catalysed by the protonation of the cytosine N3 position carried out by the glutamic acid of the amino acid motif ENV (motif VI). The formation of the covalent bond activates the cytosine C5 atom towards nucleophilic attack on the methyl group leading to the addition of the methyl group to carbon-5. The reaction cycle is closed by the elimination of the 5-position proton and the thiolmoiety, whichresolves the covalent intermediate and re-establishes aromaticity (review: Jeltsch 2002).

This description of the catalytic mechanism of DNA-(cytosine C5)-MTases by a combination of covalent catalysis and acid base catalysis is supported by a large body of experimental evidence: The covalent reaction intermediate between methylated DNA and the active site cysteine has been observed in all structures of DNA-(cytosine-C5)-MTase in complex with DNA known so far (Klimasauskas et al. 1994; Reinisch et al. 1995). In addition, the covalent intermediate has been detected biochemically with several DNA MTases including Dnmt1 and Dnmt3A (Chen et al. 1991; Hanck et al. 1993; Osterman et al. 1988; Reither et al. 2003; Santi et al. 1984; Wyszynski et al. 1993; Yoder et al. 1997) and covalent complex formation has been shown to involve the cysteine residue in the PCQ motif (Chen et al. 1991; Everett et al. 1990; Hanck et al. 1993; Reither et al. 2003).

In addition, the importance of the cysteine residue in motif IV for catalysis by prokaryotic MTases has been demonstrated by site-directed mutagenesis (Hurd et al. 1999; Wyszynski et al. 1992, 1993). The formation of a stable covalent intermediate comprising the enzyme and the target base is the basis of the efficient inhibition of DNA MTases by cytidine analogues incorporated into DNA, which currently is being investigated with respect to its therapeutic potential (review: Gowher and Jeltsch 2004).

Surprisingly, in the case of the Dnmt3A catalytic domain, the glutamic acid residue in motif VI has been shown to be very important for activity, but the removal of the active site cysteine residue did not result in a complete loss of catalytic activity (Reither et al. 2003). This finding suggests that, in addition to covalent catalysis, other mechanisms of enzyme catalysis are operative in DNA MTases (at least in the case of Dnmt3A) such as positioning of the target base and the cofactor with respect to each other and stabilisation of the transition state of methyl group transfer. In this context, it is interesting to note that Dnmt3A purified from *Escherichia coli* but also from insect cells shows only relatively low turnover rates (Aoki et al. 2001; Gowher and Jeltsch 2001; Okano et al. 1998).

This indicates that the active site of Dnmt3A is not in an ideal conformation and the cysteine residue is not ideally positioned to perform a nucleophilic attack on the C6 position. It might be possible that a covalent modification of the enzyme or an interaction with another protein could induce a conformational change of the catalytic site that activates the enzyme and switches the catalytic mechanism to the covalent catalysis scheme (Reither et al. 2003). The mammalian Dnmt1 enzyme might be a precedent for this kind of activation, because although the full-length enzyme is highly active, its catalytic domain is not active in an isolated form, which implies that an interaction of the catalytic domain with the rest of the protein is essential for the catalytic domain to adopt a catalytically competent conformation.

## Base Flipping

The first X-raystructure of a DNA-(cytosine-C5)-MTase in complex with DNA was determined with M.HhaI (Klimasauskas et al. 1994.

It demonstrated that DNA MTases completely rotate their target base out of the DNA helix prior to its methylation, a process called base flipping. After base flipping the target cytosine is no longer buried in the double helix of the DNA but is turned about its flanking sugar-phosphate bonds such that it projects out into the catalytic pocket of the enzyme. The base pairing hydrogen bonds are broken and the stacking interactions with the adjacent base pairs are lost during this process. Base flipping has been observed in all MTase-DNA complex structures known so far (Goedecke et al. 2001; Klimasauskas et al. 1994; Reinisch et al. 1995) and also in many other enzymes interacting with DNA, for example many DNA repair enzymes (reviews: Cheng and Roberts 2001; Roberts and Cheng 1998). It brings the target base into close contact to the enzyme, allowing for the intricate chemical reactions to occur and for accurate recognition of the flipped base, an important requirement for the function of DNA repair glycosylases. Also, it is a prerequisite for the catalytic mechanism as described above, because it makes the C5, C6 and N3 positions of the cytosine accessible to the enzyme.

The structure of M.HhaI is typical for all enzymes of the DNA-(cytosine-C5)-MTase family (reviews: Cheng 1995; Cheng and Roberts

2001). It comprises two domains. The larger, catalytic domain is conserved among all enzymes of this type. It consists of a central, parallel, 6-stranded â-sheet flanked by á-helices. The domain can be divided into two subdomains, one forming the binding pocket for the flipped target base, the other for the AdoMet cofactor. The structures of both subdomains are similar, and the catalytic domain most likelyarose by gene duplication (Malone et al. 1995). The smaller domain is involved in the recognition of the target sequence and structurally diverse (review: Jeltsch 2002). The only structure of a mammalian MTase catalytic domain solved so far is that of Dnmt2 (Dong et al. 2001). The protein is folded very similarly to M.HhaI; unfortunately, the reason(s) why the Dnmt2 enzyme has only a very low catalytic activity cannot be deduced from its structure.

## TARGET SEQUENCE SPECIFICITY OF MAMMALIAN DNA MTASES

All mammalian DNA MTases modify DNA at CG sites. However, the degree of specificity for the target sequence andthe preference for different methylation states of the target site varies considerably among the different enzymes.

### Specificity of Dnmt1 for Hemimethylated DNA

In 1982, Razin's group isolatedDNA MTase activity from mammalian cells that displayed a very high preference for hemimethylated CG sites (Gruenbaum et al. 1982). Later this enzyme was identified as Dnmt1, but, using oligonu-cleotide substrates, different factors for the preference of hemimethylated DNA over unmethylated were found which range from 2- to 50-fold (Fatemi et al. 2001; Flynnetal. 1996; Pradhan et al.1999; TollefsbolandHutchison1995, 1997). These differences could be due to deviations in the experimental setup (allosteric activation, see Sect. 5.1), different substrates, different sources of the proteins and different degrees of purity. For example, Bestor reported in the early1990s that treatment of Dnmt1 with proteases leads tothe loss of preference for hemimethylated target sites (Bestor 1992).

Similarly, we observed that the preference for hemimethylated DNA decreased from its original level of about 50-fold during prolonged storage of the enzyme (Fatemi et al. 2001). In the context of longer hemimethylated DNA, a 24-fold preference for ahemimethylatedtarget sitehas beendetected(Hermann et al.2004b).In vivo, this property is very important, as it enables the enzyme to copy the existing methylation pattern of the DNA after DNA replication and, therefore, to work as a maintenance MTase. The high specificity of Dnmt1 for hemimethylated target sitesisa fascinatingexample of molecular recognition, because the pres-ence of a single methyl group switches on the enzyme's activity at hemimethy-lated CG sites. The detailed mechanism of this process is not yet known.

## CG and Non-CG Methylation by Dnmt3A and Dnmt3B

Both Dnmt3A and Dnmt3B do not differentiate between unmethylated and hemimethylated substrates, and both are involved in de novo DNA methyla-tion in vivo (Gowher and Jeltsch 2001; Okano et al. 1998, 1999). Interestingly, Dnmt3A and Dnmt3B also methylate cytosine residues in a non-CG context in vitro (Aoki et al. 2001; Gowher and Jeltsch 2001; Hsieh 1999; Ramsahoye et al. 2000).

Depending on the substrate and assay system, the activity at non-CG sites varies between 0.5% and 10% of the activity observed at CG sites. In general, CA sites were found the second-best substrate for Dnmt3A and Dnmt3B. Methylation of non-CG sites by Dnmt3A has been detected also in mouse DNA (Dodge et al. 2002). However, since Dnmt1 cannot maintain this asymmetric methylation, the biological function of this activity is not known. One could speculate that non-CG methylation is important to ensure a rapid onset of a strong repression of gene expression during early embryogenesis. After some time, when additional epigenetic mechanisms like histone modification and chromatin condensation have become effective, the non-CG methylation might no longer be required.

## Flanking Sequence Preference of Mammalian DNA MTases

Another facet in the DNA interaction of mammalian DNA MTases is their flanking sequence preferences. Since it contains only two bases, the recognition sequence of these enzymes is much shorter than typical DNA interaction sites of proteins of that size, which are in the range of 8 to 14 base pairs. Therefore, it is likely that interactions between the protein and the DNA also occur outside of the central CG site, which could lead to preferences of methylation of CG sites within a certain sequence context. Such differences are usually called "flanking sequence preference" and they are conceptually distinct from the "sequence specificity", because a change in the flanks will only modify the rate of methylation, while a change in the central target site will abolish methylation.

The flanking sequence preferences of Dnmt3A and Dnmt3B have been studied in detail. Dnmt3A exhibits strong strand preference for CG sites flanked by pyrimidines and a loose consensus sequence of YNCGY (Lin et al. 2002). Later, the consensuses sequence could be refined and extended also to Dnmt3B, showing that both enzymes prefer methylation of CG sites in a RCGY context and disfavour YCGR sites (Handa and Jeltsch 2005). Interestingly, the rates of methylation of substrates differing in 4 base pairs on each site of the central CG site varied by more than more than 500fold. Comparing these numbers with the actual preference for CG over CA in a given sequence context, which is approximately 10- to 100-fold, one has to conclude that the concept of flanking sequence and central site is not fully applicable to Dnmt3A and Dnmt3B, because changes in the flanking sequence influence the reaction rate to a similar degree as a change of the central CG to CA. The flanking sequence preferences of Dnmt1 for the methylation at unmethylated CG sites have been studied as well, demonstrating the enzyme shows a clear preference for methylation within a CCGG context (R. Goyal and A. Jeltsch, in preparation).

Interestingly, a statistical analysis of human DNA methylation patterns revealed that there is a clear correlation between the average methylation level of CG sites and their flanking sequence that closely fits to the flanking sequence preferences of Dnmt3A and Dnmt3B (Handa and Jeltsch 2005). This finding demonstrates that the intrinsic preferences of Dnmt3A and Dnmt3B for certain target sites shaped the human epigenome. However, the biological implications of the sequence preferences of the Dnmt3A and Dnmt3B de novo MTases might extend even to immunology. DNA containing unmethylated CG dinucleotide sequences is immunogenic in mammals (Krieg 2002; Rui et al. 2003). In several reports it has been shown that DNA with CG flanked by purine at the 5' end and pyrimidine at the 3' end has a higher immunogenic response when compared to other sequences (Klinman et al. 1996; Krieg 2002).

This consensus sequence is identical to the high preference consensus sequence for Dnmt3A and Dnmt3B. Therefore, those flanking sequences that render high immunogenicity to unmethylated CG dinucleotide sites belong to the most preferred consensus sequence for de novo DNA MTases and hence have the lowest probability to be unmethylated in the human DNA. Thereby, the risk of an autoimmune response generated from self-DNA is minimised. This observation indicates co-evolution of de novo DNA MTases and the immune system in context with CG dinucleotides and the flanking sequences (Handa and Jeltsch 2005).

## Specificity of Dnmt2

The substrate specificity of the Dnmt2 enzyme is still not fully understood. The human enzyme has a preference for CG sites (Hermann et al. 2003) whereas *D. melanogaster* Dnmt2 was found to prefer CT and CA sites (Kunert et al. 2003). It is not clear whether or not these differences are due to the amino acid differences between both enzymes, which are only moderate. However, all these studies are hampered by the low methylation activity of the enzymes, leading to an insufficient statistical sampling. Therefore, additional experiments will be required to resolve this issue.

## Processivity of DNA Methylation by Mammalian DNA MTases

Since DNA MTases are enzymes that work on a long polymeric substrate containing several potential target sites, the processivity of the methylation reaction is an important issue for this class of enzymes. Here, processivity is defined as the preference of the enzyme to transfer more than one methyl group to one DNA molecule without release of the DNA.

## Processivity of Dnmt1

Evidence for a processive reaction mechanism of Dnmt1 dates

back to 1983 when Bestor and Ingram demonstrated that Dnmt1 methylates longer substrates faster than shorter ones (Bestor and Ingram 1983). Recently, long hemimethylated substrates were used to study the processivity of Dnmt1 in more detail using a physiological substrate. This study demonstrated that Dnmt1 modifies DNA in a highly processive reaction, and during the pro-cessive movement on the DNA it accurately copies the exiting methylation pattern (Hermann et al. 2004b). Such processive methylation of DNA implies that Dnmt1 moves along the DNA after each turnover. The mechanism of this movement is not yet clear; it might involve a sliding and a hopping process. It also is not known if Dnmt1 moves on the DNA with a directional preference.

It is tempting to speculate that the ability of Dnmt1 to methylate DNA in a processive reaction andto interact withPCNA are co-adaptations that enable the enzyme to bind to the replication fork in vivo and methylate nascent DNA immediately after DNA replication. However, its catalytic activity might not suffice to cope with the high density of CG sites in heterochromatin. Therefore, Dnmt1 might impede the progression of the replication fork if it remained tightly attached to the replication fork during replication of heterochromatic DNA. To avoid this potential complication, one could suppose that Dnmt1 is released from the replication fork during the heterochromatin replication phase, and that the methylation of heterochromatic DNA is restored after replication has taken place. This model is supported by the finding that the time gap between replication and methylation is larger for the heterochro-matic than for the euchromatic DNA (Gruenbaum et al. 1983; Leonhardt et al. 1992; Liang et al. 2002). Furthermore, it has been demonstrated that Dnmt3A and Dnmt3B also play a role in the preservation of methylation levels at heterochromatic DNA (Chen et al. 2003; Liang et al. 2002; Rhee et al. 2002).

## Processivity of Dnmt3A and Dnmt3B

Similar experiments with Dnmt3A and Dnmt3B yielded the interesting result that Dnmt3A modified DNA in a distributive reaction, but Dnmt3B was pro-cessive (Gowher and Jeltsch 2001, 2002). This was an unexpected observation because the catalytic domains of Dnmt3A and Dnmt3B are about 84% identical in amino acid sequence. However, among the 44 amino acid residues that are not identical between human and murine Dnmt3A and Dnmt3B catalytic domains, 15 include charged residues. The exchanges observed among these residues are highly biased such that, in the end, Dnmt3B carries 6 more positive charges than Dnmt3A. Therefore, Dnmt3B has a much more positively charged DNA binding cleft than Dnmt3A, which could explain why Dnmt3B methylates DNA in a processive reaction whereas Dnmt3A is distributive.

The difference in the kinetic mechanisms of the catalytic domains of Dnmt3A and Dnmt3B could be related to the distinct biological functions of these enzymes in the cell, because satellite 2 repeats (one of the major targets of Dnmt3B) are exceptionally rich in CG sites when compared with the rest of the genome (Gowher and Jeltsch2001). Dnmt3B is well suited to modify

**Dnmt3a catalytic domain** **Dnmt3b catalytic domain**

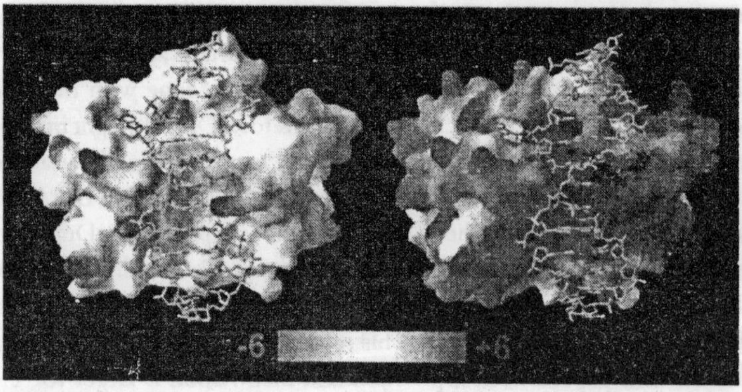


***Fig. Models of the catalytic domains of Dnmt3A and Dnmt3B. The models were prepared using M.*****Hha*****I as template as described in (Gowher and Jeltsch 2002). The surface of the proteins was coloured according to the electrostatic potential calculated using Swiss PDB viewer version 3.7.b2. To illustrate the location of the DNA binding cleft in the enzymes, the DNA as seen in the M.*****Hha*****I-DNA complex is shown inorange, the AdoMet is shown in* green**

these regions, because after targeting to the DNA it can methylate several cy-tosine residues in a processive reaction. The distributive reaction mechanism of Dnmt3A might explain why it cannot replace Dnmt3B at satellite repeats in vivo, although the Dnmt3A enzyme can methylate these regions.

So if the processive mechanism has such obvious advantages, why did Nature invent distributive enzymes like Dnmt3A? One advantage of a distributive enzyme could be that its activity is under better control, because it has to be directed to the DNA for each single methylation event. Therefore, a distributive enzyme depends on a mechanism targeting it to the sites of action much more so than a processive enzyme, where one targeting event will lead to the transfer of several methyl groups to the DNA.

In line with these considerations, Dnmt3A has been associated with the methylation of single-copy genes and retrotransposons (Bourc'his and Bestor 2001, 2004; Hata et al. 2002) and it is critical to the establishment of the genomic imprint during germ cell development (Kaneda et al. 2004). Therefore, Dnmt3A is involved in the methylation of defined target sites, whereas Dnmt3B (at least as far as the methylation of heterochromatic repeats is concerned) catalyses the complete methylation of large DNA domains. One could envisage that Dnmt3A contacts a targeting factor and thereby keeps indirect contact (via the targeting protein) to the DNA. This mechanism would allow for efficient methylation of the DNA at sites that are determined by the specificity of the targeting complex.

## CONTROL OF DNA MTASE ACTIVITY IN MAMMALIAN SYSTEMS

The mechanism by which mammalian DNA MTases create a specific DNA methylation pattern that carries additional information is one of the most fascinating questions regarding the function of these enzymes. Although the exact mechanism of pattern generation is not certain, it clearly depends on the control of the enzyme's activity by different instances that include control of gene transcription, covalent modification and interaction with regulatory proteins. The transcriptional controlofmammalian Dnmts has been reviewed recently (Pradhan and Esteve 2003b) and is beyond the scope of this review, which focuses on enzymology. Dnmt1 isolated from mammalian cell lines has been shown to carry some phosphoryl groups (Glickman et al. 1997).

However, the functional relevance of this modification is not yet known, and it is not clear if post-translational modifications occur with

Dnmt3A, Dnmt3B or Dnmt2 as well. In the following paragraphs the interactions of MTases with regulatory proteins will be discussed.

**Allosteric Activation of Dnmt1**

Surprisingly, t he isolated catalytic domain of Dnmt1 is not catalytically active, although it contains all the amino acid motifs characteristic for cytosine-C5 MTases (Fatemi et al. 2001; Margot et al. 2000; Zimmermann et al. 1997). These results demonstrate that the N-terminal part of Dnmt1 has an important role in controlling the activity of the protein, such that Dnmt1's N-terminal part could be considered a "regulatory protein". A similar observation was already made by Bestor (1992) by demonstrating that a proteolytic cleavage of Dnmt1 just between the catalytic domain and the N-terminal domain leads to a strongly increased activity of Dnmt1 towards unmethylated target sites (Bestor 1992). In this study, the C- and N-terminal parts of Dnmt1 most likely remained in contact, but the proteolytic cleavage induced a conformational change that activated the enzyme.

Interestingly, Dnmt1 bears at least two separate DNA binding sites, at least one in the N-terminal part and one in the C-terminal part (Araujo et al. 2001; Fatemi et al. 2001; Flynn and Reich 1998). The enzyme can interact with its target DNA and, in addition, with a second DNA molecule that functions as an allosteric regulator. Binding to methylated DNA activates Dnmt1 for methylation of unmodified target sites (Bacolla et al. 1999; Fatemi et al. 2002; Fatemi et al. 2001). Steady-state kinetic experiments demonstrate that the N-terminal part of Dnmt1 has a repressive function on the catalytic domain, which is relieved after binding of methylated DNA to the N-terminus (Bacolla et al. 2001).

Experimental evidence suggests that binding of methylated DNA occurs within the Zinc-domain, which forms a direct protein/protein contact to the catalytic domain of the enzyme (Fatemi et al. 2001) or to a short motif in between the PCNA interaction site and the nuclear localisation signal (NLS) (Pradhan and Esteve 2003a). Given these results, at least three different states of Dnmt1 can be distinguished: The isolated catalytic domain is inactive towards hemimethylated and unmethylated DNA. With unmethylated DNA the full-length enzyme shows low activity. In the presence of methylated DNA, the activity of

Dnmt1 is much higher, suggesting that the N-terminal part has two effects:

(1) It stimulates the C-terminal part for general activity and

(2) either unmethylated DNA binding to the N-terminal part inhibits the enzyme or binding of methylated DNA stimulates the enzyme, leading to an increased methylation of unmodified sites.

This allosteric activation is a surprising effect, as it means that, in the presence of methylated DNA, Dnmt1 loses specificity for hemimethylated DNA and also starts working as a de novo MTase. Therefore, activated Dnmt1 is less accurate in copying an existing methylation pattern, which at first sight appears as a mis-adaptation for a maintenance MTase. After allosteric stimulation, Dnmt1 has a similar activity on unmethylated and hemimethy-lated DNA, suggesting that this enzyme could also have a role in de novo methylation of DNA. Activated Dnmt1 could support Dnmt3A and Dnmt3B in de novo methylation, a conclusion that is in agreement with in vivo data demonstrating Dnmt1 is required for de novo methylation (Liang et al. 2002) and overexpression of Dnmt1 can cause de novo methylation of DNA (Bin-iszkiewicz et al. 2002). This assumption is also supported by the finding that Dnmt1 and Dnmt3A interact with each other (Datta et al. 2003; Kim et al. 2002).

The allosteric activation mechanism of Dnmt1 makes DNA methylation behave in an all-or-none fashion, because some methylation will always attract more methylation. In addition, epigenetic signalling comprises several positive feedback loops: Initial DNA methylation could induce histone 3 ly-sine 9 methylation or histone deacetylation (Cameron et al. 1999; Fahrner et al. 2002; Sarraf and Stancheva 2004; Tariq et al. 2003). These responses in turn could trigger additional DNA methylation (Bachman et al. 2003; Jackson et al. 2002; Lehnertz et al. 2003; Tamaru and Selker 2001). Furthermore, methylation of DNA could attract MeCP2 that itself would target Dnmt1 to the DNA (Kimura and Shiota 2003).

Therefore, in a steady-state situation only completely unmethylated and fully methylated regions of the DNA coexist, which are separated by chromatin boundary elements. This all-or-none behaviour might increase the efficiency of epigenetic circuits in switching on and off gene expression. These mechanisms also explain the observation that methylation tends to spread from heavily methylated regions of the DNA into neighbouring unmethylated regions, which is often observed in cancer cells.

## Stimulation of Dnmt3A and Dnmt3B by Dnmt3L

De novo methylation by Dnmt3A and Dnmt3B is regulated by at least one additional protein, namely Dnmt3L, which shows clear homology to the Dnmt3A and 3B enzymes (Aapola et al. 2000). However, Dnmt3L carries mutations within all conserved DNA-(cytosine-C5)-MTase motifs. This observation suggests that Dnmt3L adopts the typical MTase fold, but it does not have catalytic activity. In co-transfection experiments, Dnmt3L has been shown to stimulate DNA methylation by Dnmt3A in human cell lines (Chedin et al. 2002). In vitro studies demonstrated an approximately 15-fold activation of Dnmt3A and Dnmt3B by Dnmt3L (Gowher et al. 2005). Biochemical studies demonstrate Dnmt3L directly interacts with Dnmt3A and Dnmt3B via its C-terminal domain (Gowher et al. 2005; Hata et al. 2002; Suetake et al. 2004) and induces a conformational change of Dnmt3A that facilitates DNA and AdoMet binding. However, the interaction of Dnmt3A and Dnmt3L is transient, and Dnmt3L dissociates from Dnmt3A-DNA complexes. Therefore, Dnmt3L acts as a substrate and coenzyme exchange factor on Dnmt3A and Dnmt3B (Gowher et al. 2005).

Dnmt3L is expressed during gametogenesis and embryonic stages (Bour-c'his and Bestor 2004; Bourc'his et al. 2001), showing a similar expression pattern as the Dnmt3A and Dnmt3B enzymes. Dnmt3L knock-out mice display a normal phenotype (Bourc'his and Bestor 2004; Bourc'his et al. 2001; Hata et al. 2002). Homozygous female mice are fertile, but when crossed with wild-type males their pups die at embryonic day 10.5. Analysis of the DNA methylation pattern showed that the female imprint was not properly established in oocytes of Dnmt3L knock-out females (Bourc'his et al. 2001; Hata et al. 2002). Homozygous male knock-out animals are sterile because of defects in spermatogenesis. Methylation analysis showed major loss of methylation in spermatogonial stem cells, leading to male infertility (Bourc'his and Bestor 2004; Hata et al. 2002). These strong phenotypes of Dnmt3L knock-out mice illustrate the importance of the stimulatory effect of Dnmt3L on Dnmt3A and Dnmt3B in vivo. It is to be expected that more regulators (inhibitors and stimulators) of Dnmts will be discovered in the future.

## Future Perspectives

The cellular memory of developmental decisions is crucial in the development and maintenance of multicellular organisms. Failure in the propagation of the cellular memory of differentiated states is a major reason for cancer and other diseases. Cellular memory is mediated by epigenetic switches including DNA methylation in mammals. DNA MTases, the enzymes that set up the pattern of DNA modification and thereby impose additional information on the DNA, are central actors in epigenetic information transfer. However, many mechanistic features of these fascinating enzymes are incompletely characterised so far. Future biochemical experiments will address issues like substrate specificity, reaction mechanism, control of enzyme activity, targeting of methylation to certain DNA regions and interaction of MTases with other proteins involved in epigenetic processes. A more detailed understanding of the behaviour of DNA MTases shall enable us to get a better grasp of epigenetic regulation as a whole.

# 8

# Familial Hydatidiform Molar Pregnancy

Imprinting is the uniparental expression of a set of genes. Somatic cells carry two haploid sets of chromosomes, one maternal and one paternal, while germ cells contain only one of the two forms of chromosomes, male or female. This implies that during early embryogenesis the cells committed for developing the future germ cell lineage, the primordial germ cells, which are diploid, have to undergo a total chromosome reprogramming process.

This process is delicately controlled during gametogenesis to ensure that males and females have only their respective form of gametes. The machinery involved in this process is yet poorly defined. Familial hydatidiform molar (HM) pregnancy is an abnormal form of pregnancy characterized by hydropic degeneration of placental villi and abnormal, or absence of, embryonic development. To date, the molecular defect causing this condition is unknown. However, in a few studied cases, the presence of paternal methylation patterns on the maternal chromosomes was observed. In this chapter, we summarize what is known about methylation aberrations in HMs and examine more closely the proposed hypothesis of a maternal germline imprinting defect.

## INTRODUCTION: THE LIFE CYCLE OF AN IMPRINT

In the process of fertilization, both male and female gametes contribute equal amounts of genetic material to the newly formed zygote; however, the two haploid genomes (in the gametes) are not functionally equal (Walter and Reik 2001; Ferguson-Smith and Surani

2001). A set of genes is marked for silencingof transcription in one of the gametes but transcribed from the other. These sets of marked genes are said to be imprinted. Imprinting in somatic tissues is defined as mono-allelic transcription of a given gene depending on the parental origin of the chromosome. The imprinting process defines.

One unique example in humans for a disease that is manifested, or caused, by an imprinting defect is recurrent familial hydatidiform moles (HMs) (OMIM 231090). HMs mimic uni-parentalmouse embryos (Barton et al. 1984) where androgenotes develop normal extra-embryonic tissues but there is no or little embryonic development, while parthenogenotes, on the other hand, give rise to the opposite phenomenon, normal embryonic development with poor development of extra-embryonic tissues. The exact molecular mechanism leading to familialHM is currently unknown. In this chapter, we will discuss the reasons that led investigators to suggest that it is a maternal germline defect in establishing the maternal imprinting marks and the validity of this hypothesis.

## FAMILIAL HYDATIDIFORM MOLAR PREGNANCY

### Diagnosis and Clinical Manifestations of Molar Pregnancies

HM is an abnormal form of human pregnancy characterized by hydropic degeneration of placental villi with the absence of, or abnormal, embryonic development. Based on the histology of the evacuated molar tissues, HMs are divided into two types: complete hydatidiform moles (CHMs) and partial hy-datidiform moles (PHMs). CHMs are characterized by hydropic degeneration of all villi and absence of embryo, cord, and amniotic membranes. All the villi are:

(1) enlarged with cisternae,

(2) avascular, and

(3) surrounded by areas of trophoblastic proliferation.

PHMs are characterized by focal trophoblas-tic proliferation with a mixture of normal-sized villi and edematous villi. The trophoblastic proliferation is less pronounced than in complete moles. An embryo, cord, and amniotic membranes are usually present in partial moles (Copeland 1993; Bonilla-Musoles 1993). This subdivision is supported by karyotype data, which show that most complete moles are diploids while partial moles are triploids. We note that moles are not always easily divisible into partial and complete moles; in a minority of cases, embryonic tissues are found in complete moles evacuated at early stages (Zaragoza et al. 1997; Fukunaga 2000) and some partial moles are found diploid with biparental origin.

## Epidemiology and Genetics of Molar Pregnancies

The most recent reports estimate that 80% of CHMs have a diploid genome and are androgenetic. Among those, 60% are monospermic and 20% are dispermic (Kovacs et al. 1991; Lindor et al. 1992). The remaining 20% have a biparental genomic contribution to their genome. Most reported cases of HMs are sporadic and not recurrent. Occasionally, recurrent cases have been reported in one family member (Patek and Johnson 1978; Neumann 1980; Thavarasah and Kanagalingam 1988; Narayan et al. 1992; Tuncer et al. 1999; Ozalp et al. 2001; Fisher et al. 2000) and in a few cases, in at least two related women (familial cases) (Ambani et al. 1980; La Vecchia 1981, Parazzini et al. 1984, Seoud et al. 1995; Kircheisen and Schroeder-Kurth 1991; Sensi et al. 2000; Judson et al. 2002; Fisher et al. 2002; Al-Hussaini et al. 2003; Hodges et al. 2003; Fallahian et al. 2003; Agarwal et al. 2004; for review see Fisher et al. 2004). In several of these cases, women with recurrent moles had also abortions at various gestational stages and some achieved normal pregnancies and gave birth to healthy babies (Ambani et al. 1980; Seoud et al. 1995; Fallahian et al. 2003).

Consanguineous marriages were observed in many of these families, and in all of them the segregation of the defect is compatible with an autosomal recessivemodeoftransmission, with thewomenhavingrecurrent moles being homozygous for the defective locus.

One group characterized the parental contribution to familial moles and demonstrated, using homozygosity mapping, that a maternal locus mapping to 19q13.4 between markers D19S924 and D19S890 is responsible for this condition (Moglabey et al. 1999). This locus was confirmed by other groups and on several families that allowed narrowing down the candidate region to 1.1 Mb flanked bymarkers D19S418 and AAAT11138 (Sensi et al. 2000; Hodges et al. 2003). However, not all families show linkage to 19q13.4, indicating the genetic heterogeneity of this disease (Judson et al. 2002; Slim et al. 2005), which could also reflect heterogeneity in the molecular mechanisms leading to familial moles.

## Methylation Analysis in Molar Tissues

Methylation of DNA at the cytosines' fifth carbon is the most abundant modification of DNA in the human genome. This fifth base (5-methyl-cytosine: 5mC) occurs at a frequency of about 3%–4% of total cytosines. Most 5mCs occur at clusters called CpG islands. These are found in the promoter region of about one-third of human genes. These CpG islands play an important role in the regulation of gene activity and expression of the nearby genes. Together with other epigenetic signals such as histone acetylation/methylation, they impose an open or closed chromatin structure that is associated with expressed (on) or repressed (off) gene expression. Regions that are actively transcribed (euchromatin) have promoter regions with mostly unmethylated CpG sites, acetylated histone tails and methylated lysine 4 on H3 histone sub-units, while transcriptionally silent regions (heterochromatin) have mostly methylated CpG sites, deacetylated histones and methylated lysine 9 on H3 subunits (Fournier et al. 2002; Tamaru and Selker 2001). Imprinted genes that make the asymmetry in gene expression between the two sets of male and female gametes, and thus the two parental sets of chromosomes, are associated with differentially methylated regions (DMRs). These DMRs are CpG-rich regions that are heavily methylated on the non-expressed (repressed) allele and nearly devoid of methylation on the expressed allele.

The importance of correct methylation settings in the gametes and early embryogenesis is illustrated by the facts that aborted cloned animals (following nuclear transfer) show irregular methylation patterns (Kang et al. 2001; Beaujean 2004; Chen et al. 2004; Jaenisch 2004). Low methylation levels in sperm were also found to give lower rate of pregnancy in assisted reproductive techniques (Benchaib et al. 2005). Molar pregnancies—whether andro-genetic or biparental (sporadic or familial)—are identical at the histopatho-logical level; the only known functional difference between the maternal and the paternal genome is in the expression of imprinted genes. This has led to a common belief that imprinted genes play an important role in the pathology of moles and that a defective gene causing their deregulation could underlie the etiology of familial biparental molar pregnancies.

The above hypothesis was first tested by Judson et al. (2002) who studied a single biparental molar tissue from a family with recurrent HM. In this study, the authors made a detailed analysis of a well-characterized set of DMRs associated with *H19*, *KCNQ1OT1* (*LIT1*) *SNRPN*, *PEG1*, *PEG3*, and four with the *GNAS1* locus. They showed that seven of the nine analyzed maternally methylated DMRs (at *KCNQ1OT1*, *SNRPN*, *PEG1*, *PEG3*, and two of the four *GNAS*) were not methylated.

For the paternally methylated DMRs, again, not all of them behaved similarly; the *H19* DMR had a normal methylation level while the *NESP55* DMR (at the *GNAS1* locus) was completely hypermethylated, indicating that the maternal allele behaved like a paternal allele. In the above study, no DNA polymorphisms were used to track the parental origin of the abnormally methylated alleles in the molar tissue. Indeed, this is needed to identify the parental alleles and see whether abnormal methylation is affecting both of them or only one. Abnormal methylation at both parental alleles would indicate epigenetic changes during the proliferation of the trophoblast; while an abnormal methylation exclusively on the maternal alleles may indicate a primary defect that could be traced in origin to the maternal defect leading to familial moles.

We also analyzedthe methylation offour DMRs, twopaternally methylated, *H19* and *NESP55*, and two maternally methylated, *SNRPN* and *PEG3*, in two molar samples from one family (El-Maarri et al. 2003). Using a quantitative method (El-Maarri et al. 2002, 2004), we found similar trends of abnormal methylation like the ones reported by Judson et al..

The studied paternal methylation (at *H19* and *NESP55*) in the two molar samples [biparental complete hydatidiform moles (BiCHM) 9 and 16] were lower than that of androgenetic complete hydatidiform moles (AnCHMs) and higher than that of normal chorionic villi and total blood DNAs, while the maternal methyla-tion (*SNRPN* and *PEG3*) were decreased. This suggests that portions of the maternal chromosomes are assuming a paternal methylation patterns. To investigate whether the two parentalalleles are affected by the abnormal DNA methylation, we looked for single nucleotidepolymorphisms (SNPs) and identified informative ones in a number of DMRs in one or two molar tissues (*SNRPN* in BiCHM16; *NESP55* in *H19* in both BiCHM9 and BiCHM16).

Bisulfite sequencing of individual clones at these DMRs showed paternal methylation pattern most maternal chromosomes; *H19* acquired methylation marks while *SNRPN* did not show methylation as it should on the maternal allele.

This partial shift from the maternal to paternal patterns of methylation is intriguing and deserves to be investigated on additional imprinted genes. In case a similar shift is observed on all imprinted genes, this would indicate an abnormality in the setting or maintaining of the correct maternal methylation imprinting marks on the maternal chromosomes rather than a general failure in the methylation machinery. This is further supported by the fact that the two patients with recurrent HMs have both normal patterns and levels of methylation at the same four imprinted loci in their blood (El-Maarri et al. 2005).

## IMPRINTED GENE EXPRESSION ANALYSIS

Transcription analysis of imprinted genes in sporadic androgenetic moles showed abnormal imprinted gene expression and relaxation of imprinting in some androgenetic moles (Ohlsson et al. 1999; Ariel et al. 2000; Kim et al. 2003).

These results are compatible with our data on androgenetic moles, in which we observed at *H*I *9* a lower level of methylation than that observed in sperm DNAs. In familial biparental moles, only one study addressed the expression of one maternally expressed gene, $p57^{KIP2}$ *(CDNK1C;* Fisher et al. 2002).

The authors used mouse monoclonal antibody against the $p57^{KIP2}$ protein on histo-logical sections from familial and sporadic molar tissues. They demonstrated that $p57^{KIP2}$, which is expressed in normal first trimester placenta, is not expressed in biparental moles (familial and sporadic) nor in androgenetic moles.

## Hypothesis of a Germline Imprinting Reprogramming Defect

Familial biparental HM pregnancy could be regarded as a disease of imprint reprogramming that takes place in the affected females to produce female gametes with paternal methylation imprints. However, to date there is no direct proof for such hypothesis mainly because of the impossibility of studying germ cells from such patients. Hereon, we list the reasons/observations that support such a hypothesis as well as reasons against it.

As indirect support for the germline imprinting defect hypothesis involving the maternal chromosomes we could list:

(1) the fact that at both gross morphology and histology levels both familial biparental moles and andro-genic moles are undistinguishable;

(2) the similarity in the pattern of growth between biparental or androgenetic moles with that of experimentally created mouse androgenotes with two male pronuclei;

(3) methylation analysis of the few available molar samples revealing that differentially methylated regions associated with imprinted genes show variable degree of paternal methyla-tion patterns only on the maternal alleles;

(4) the fact that only methylation at imprinted loci seem to be affected [the analysis of two X-linked genes (Judson et al. 2002; El-Maarri et al. 2003) revealed that they are normally methylated].

Reasons that could argue against a maternal germline imprinting defect are: first, all performed studies on molar tissues were done on samples of 6–14 weeks of gestation in which several changes could have occurred since fertilization, mainly because of the dynamic nature of early trophoblast and the postzygotic methylation changes that take place between fertilization and implantation; second, molar pregnancies are benign tumors of the tro-phoblast, and several studies have shown gain or loss of methylation marks at several imprinted genes including *PEG3*, *PEG1*, *SNRPN*, and *H19* in a variety of tumors.

In colorectal cancer and Wilms' tumors, a similar shift from a maternal methylation pattern to a paternal one was observed at the

*H19-IGF2* imprinted region (Steenman et al. 1994; Moulton et al. 1994; Taniguchi et al. 1995; Maegawa et al. 2001; Cui et al. 2001; Nakagawa et al. 2001); third, studies on sporadic, (androgenetic and biparental) moles demonstrated abnormal methylation or/and expression of a number of non-imprinted genes including oncogenes, tumor suppressors, and genes involved in protein synthesis, cell cycle, and intercellular communication (Olvera et al. 2001; Kato et al. 2002; for review see Li et al. 2002; Batorfi et al. 2003; Durand et al. 2003; Xue et al. 2004). It would be expected that at least some of these genes are also deregulated in familial biparental moles. The presence of normal methylation levels on two X-linked genes in familial biparental moles does not allow reaching a conclusion on the methylation status of non-imprinted genes. A more comprehensive analysis of a large number of non-imprinted genes in molar tissues is needed.

## Variability of Phenotype

One important observation derived from the methylation analysis on MoLb1 is that the abnormalities in the level of methylation was not the same at all loci and in all samples. This may also be true in other cases but could not be seen since only one molar tissue was analyzed (Judson et al. 2002). This also may be allelic and restricted to some families where variability in the phenotype of the conceptuses of these patients ranged from complete moles to spontaneous abortions at various developmental stages, and normal birth. This variability could be explained by the contribution of other environmental or/and genetic factors to the disease phenotype.

## Concluding Remarks

Familial HM pregnancy is manifested by abnormal imprinting methylation marks. This abnormal pregnancy reflects the importance of establishing and maintaining the correct methylation marks for normal embryogenesis. The gene defect underlying this disorder is still to be identified; when identified it will increase our understanding of the protein machinery involved in the setting and maintenance of imprinting during embryogenesis andwill answer the question as to when and how this abnormal paternal methylation was acquired in these tissues.

# 9

# Dual Inheritance

Genetic inheritance in higher organisms normally refers to the transmission of information from one generation to the next. Nevertheless, there is also inheritance in somatic cells, characterised by the phenotypic stability of differentiated cells that divide (such as fibroblasts and lymphocytes), and also mitosis of stem line cells, which gives rise to another stem line daughter cell, and one that will differentiate. Thus, there is a dual inheritance systems in these organisms, one of which is genetic and the other epigenetic.

In thelatter, heritable information is superimposedon DNA sequences, and one well-known mechanism is heritable methylation of cytosine. Much information will come from the human epigenome project that will reveal the patterns of DNA methylation in distinct differentiated cells. There have also been innumerable studies on the abnormal de novo methylation and silencing of tumour suppressor genes in cancer cells.

## INTRODUCTION

In 1990 John Maynard Smith published a paper "Models of a dual inheritance system". In his Introduction he wrote:

In higher plants, animals and fungi there are two inheritance systems, as follows:

1. The familiar system, depending on DNA sequence, used in transmitting information between sexual generations.
2. An epigenetic inheritance system (EIS), responsible for cellular inheritance during ontogeny—for example fibroblasts give rise

> to fi-broblasts, epithelial cells to epithelial cells, and *Drosophila* wing discs continue to be wing discs in serial transfer.

This paper was in response to the proposals by Jablonka and Lamb (1989) that epigenetic changes might sometimes be transmitted by sexual reproduction, following earlier discussion of the same theme (Holliday 1987). These proposals were further elaborated in their book *Epigenetic Inheritance and Evolution* (Jablonka and Lamb 1995). It was characteristic of Maynard Smith that he immediately recognised the significance of the epigenetic system, and it was he who first coined the term "dual inheritance".

In the standard literature it is not usual to categorise the division of the specialised cells of higher organisms as an inheritance system. Instead, it is simply stated that some differentiated cells (e.g. lymphocytes and fibroblasts) are capable of mitotic division, whereas others (e.g. neurons and muscle cells) are not. Traditionally, genetics and inheritance in multicellular organisms refers only to sexual reproduction. However, in micro-organisms such as yeasts and fungi, it is common to refer to asexual or vegetative reproduction, and there may also be a parasexual cycle. Thus, in a microbial eukaryote an induced mutation will be inherited through mitotic division to form a clone. The same mutation can also be transmitted through meiosis, or segregate from a diploid nucleus.

Historically, it was probably the work of Hadorn and his colleagues that first demonstrated the importance of somatic cell inheritance (reviewed in Ursprung and Nothiger 1972). In their experiments (included by Maynard Smith under system ii), imaginal disc tissue of *Drosophila* is inherited in a *determined* state. The cells are undifferentiated, but when the tissue is treated with the hormone ecdysone they differentiate. Thus, leg disc tissue differentiates into recognisable leg structures, wing tissue into wing and so on. The determined but undifferentiated disc tissue can be propagated in the abdominal cavity of adult flies, in some cases for hundreds of generations. Sometimes the determined state changes into that of a different tissue. This is known as transdetermination, and it is not random, but follows certain rules, such that determined state A can change to B or C, but not to D. However, B or C might change to D. The frequencies of transdetermination also vary, and are in fact an inherited property of the particular determined state. There is nothing intrinsically different between the inheritance of determined states to the inheritance of differentiated states.

The phenotype of a fibroblast is the result of the specialised expression of one set of genes, producing so-called luxury proteins, and the lack of expression of all those genes that produce luxury proteins in other specialised cells. This phenotype is stably inherited through serial subculture—and also in vivo—until the cells become senescent and post-mitotic. It is well established that the pheno-types of specialised cells are very stable, and what would be the equivalent of transdetermination does not occur. It is generally assumed that the change of one specialised cell type into another never occurs, but one should be careful not to make this a dogmatic assertion, because in developing or adult organisms it might happen in specific circumstances, for example, in limb or tail regeneration.

## The Inheritance of DNA Methylation

It is important to understand that the classical gene regulatory systems categorised in bacteria do not provide an inheritance system. If a gene is expressed in response to a specific inducing chemical in the medium, the daughter cells produced by division will only have the same phenotype provided the inducer is present. If the inducer is removed then the cell reverts to its original state. The phenotype of a differentiated eukaryotic cell is not dependent on extracellular signals. Cells such as fibroblasts require protein factors in serum in the medium in order to grow, but these same factors are not necessary for the cellular phenotype. Therefore, there are intrinsic mechanisms that maintain the phenotype.

Recognising this, I proposed with my colleague John Pugh (Holliday and Pugh 1975) specific mechanisms involving the modification of DNA. There were four features in the development of higher organisms that were emphasised:

1. The modification of a controlling or regulatory region of DNA adjacent to a gene can occur. The modification would silence a gene, and in the absence of modification the gene would be active, or vice versa.

2. The modified and non-modified forms of the gene would be stably inherited.
3. There would be switching between modified and non-modified states, either during normal development, or in stem cell situations. In the latter case, a stem cell would produce a daughter the same as itself and one that was destined or determined to differentiate subsequently. In the former case, a cell A could give rise to two B cells, with new properties, or to two different cells, B and C.
4. There would be developmental clocks capable of counting cell divisions, and at the end of a specified number of cell divisions a regulatory mechanism of some kind would be triggered.

We proposed that the modification could be based on the methylation of cytosine in DNA to form 5-methyl cytosine, and also that the methylation pattern could be inherited if there was a maintenance methylase that recognised hemi-methylated DNA at the replication fork and methylated the new strand. This enzyme would not recognise non-methylated DNA. We also proposed, followingScarano (1971), that cytosine might be deaminated atspecific sites to form thymidine, or that adenine would be converted to guanine. In the same year, Riggs (1975) proposed essentially the same DNA methylation mechanism and applied it in the context of the inactivation of one X chromosome in female mammals.

The two X chromosomes are in the same cellular milieu, yet one is very stably maintained in an active form and the other in an inactive form. His model also involved a rapid initial switch in which one chromosome was marked by methylation, and this process immediately inhibited the methylation of the other X chromosome. Also in 1975, Sager and Kitchin suggested that the methylation of DNA may control the processes of elimination or inactivation of chromosomes in various contexts. From the many studies of bacterial methylation, they suggested that non-methylated DNA might be lost through the activity of restriction enzymes, or that genes might be inactivated. Our models for the events listed above were based on cell lineages, which is probably incorrect, because many developmental events are known to occur in groups of cells, which Crick and Lawrence (1975) dubbed "polyclones" to describe the compartments in *Drosophila* development. Subsequent to 1975, a vast amount of evidence has accumulated that differential inherited methylation does occur in higher organisms and that the methylation of CpG islands near promoter sequences silences genes (Millar et al. 2003; Beck and Olek 2003). On the other hand, evidence that there are specific base changes in DNA has not been forthcoming, except in the context of antibody gene variability (Neuberger et al. 2003; Pham et al. 2003). In 1987, I adopted Wa ddington's use of the word epigenetic, which I took to mean the totality of mechanisms that are required to unfold the genetic programme for the development of a complex organism (Holliday 1987).

In particular, I discussed epigenetic defects that were changes in gene expression following methylation or demethylation, and suggested that these might be an important contributor to ageing, and also, following an earlier proposal (Holliday 1979), that they might be responsible for changes in gene expression during tumour progression. There is now a huge literature documenting the methy-lation and silencing of tumour suppressor genes in many types of cancer (Jones and Baylin 2002). In contrast, the role of DNA methylation changes in normal development is not well documented. Although there are many suggestive observations (nine were listed in Holliday 1996), it would be true to say that most developmental biologists do not regard DNA methylation as a key mechanism in development.

Nor is there widespread acceptance of the proposal that the switching on or off of genes coding for luxury proteins is based on DNA methylation. It is more commonly believed that the modification of the histones of chromatin—for example by acetylation, de-acetylation or methylation—is a more likely mechanism. Although this may provide a means whereby the formation of inactive heterochromatin at CpG islands occurs in non-dividing cells, it does not by itself provide a mechanism for the strict heritability of a given cell phenotype. Those who work on *Drosophila* cite the behaviour of the Polycomb group of proteins that appear to provide a basis for heritability of given chromatin configurations, at least over the fairly small number of generations that occur during fly development (Grewal and Maozed. 2003; Lavigne et al. 2004). It is not known whether these proteins could also explain the heritability of the determined state of imaginal disc cells during long-term serial passaging.

## Mutations and Epimutations

In the early days of mammalian somatic cell genetics there was a controversy between those who believed the cells could be handled in much the same way as micro-organisms, and those who believed that mammalian cells' genetic behaviour did not correspond to that of simple eukaryotes, and that their variability was not just due to simple mutations. As is often the case in controversies, both viewpoints proved to be correct (Holliday 1991). Experiments with CHO (Chinese hamster ovary) cells provided strong evidence that mutations could be induced in many housekeeping genes; these could be shown to be recessive in hybrids and to reappear in segregants.

It was proposed that CHO cells were functionally hemizygous, meaning that substantial parts of the genome were haploid, and this facilitated the isolation of mutants. Later it was shown that this haploidy was due in many cases to the presence of a silent gene on the homologous chromosome. The silent gene was methylated and could be reactivated by the powerful demethylating agent 5-azacytidine. In some cases bothgenes were inactivated by methylation. In fact, it became clear that if there was no selective disadvantage during the laboratory growth of CHO cells, any gene might become inactivated through de novo methylation.

These studies provide clear evidence for a dual inheritance system, but not one which involves sexual reproduction. Instead, we have classical mutation, induced by mutagens, and involving a change in DNA base sequence, in which there is great stability and rare back-mutation. In contrast, we have gene inac-tivation and activation due to alterations in DNA methylation. The new term epimutation was coined.

CHO cells are transformed, and it is in such cells that uncontrolled changes in DNA methylation can occur. It is extremely unlikely that similar changes occur in normal diploid cells, except perhaps at very low frequency. It is appropriate to state that tumour progression involves epimutations, and it is probable that early steps in tumourigenesis result in both chromosomal destabilisation and loss of normal methylation control. Thus, genetic and epigenetic instability provides the variability on which cellular selection can act, and this ultimately leads to malignancy. Epimutations play no part in development, but they may well be important during ageing.

This would introduce "random noise" in the normal controls of gene regulation, which would contribute to the overall processes of ageing. Insufficient information is available to know whether this is more or less important than other likely contributors to the senescent phenotype, such as mitochondrial deletions, classical chromosomal mutations, chromosome abnormalities, the accumulation of defective proteins, membrane defects and so on. One promising approach, which has not been exploited, would be to examine the frequency of ectopic expression of a luxury protein in differentiated ageing cells that do not normally synthesise that protein.

## Epigenetic and Classical Inheritance

The successful study of classical genetics in any organism depends on the variability in phenotype produced by gene mutations. In most cases the frequency of mutations is increased by the application of mutagens.

The mutations, which may be dominant, recessive, intermediate or neutral, are due to changes in the base sequence of DNA, whether base substitution, addition, deletion, inversion or a small chromosomal change. These mutations are normally very stable and are faithfully transmitted through meiosis. Larger chromosome changes are not usually regarded as mutations, and they may be unstable at meiosis if they disrupt homologous pairing and/or segregation.

Classical genetics is based on clones and lineages. The mammalian zygote forms a clone in which all the cells have the same genotype, except in special casessuch as the DNA rearrangementsin theassemblyof antibodygenes. Also, in female mammals random X chromosome inactivation produces a mosaic of two types of cell within thebody. (Clonal gene expression may then be seen, as in the tortoiseshell cat.) In the formation of the sperm and eggs, complete haploid genomes are derived from the diploid germ cells, and recombination with chromosome re-assortment ensures that every gamete is genetically distinct. It is generally accepted that the influence of the environment on these events is nil, or minimal.

Epigenetic inheritance is very different. The development of the zygote soon leads to the segregation of gene activities, so the cells diverge in their phenotypes. They can be said to acquire different "epigenotypes". They all have the same genes, but their pattern of activities becomes very different. Development is not clonal, because groups of cells may follow the same developmental pathway and have the same developmental fate. Thus, certain groups will form muscle, the central nervous system and so on. There are also developmental signals from given cells that influence other cells. Both the transmitting and the receiving cells are not behaving clonally because they may form part of a group. In a stem cell situation, epigenetic switches continually produce cells with new epigenotypes. Large populations of cells may all be behaving the same way, but differently from other types of stem cells.

Genomic imprinting comprises a set of epigenetic signals, superimposed on the DNA sequences, and DNA methylation is strongly implicated. The imprints in the male and female gametes complement each other to produce normal development. Although some imprints may be long lasting, it is quite likely that some are lost during the growth of somatic tissues. This may be the reason why the cloning of animals using somatic cell nuclei is usually unsuccessful, or the cloned animal is defective. Imprints are erased during gametogenesis, andnew ones are imposed. It is likely that any other epigenetic signals, acquired for instance in germ line cells, are also erased. The fate of epigenetic defects is contentious.

It was suggested that the loss of normal methylation can be repaired by recombination at meiosis, as heteroduplex DNA will be hemimethylated and recognised by the maintenance methylase (Holliday 1987). In the fungus *Ascobolus immersus*, it has been possible to follow the inheritance of a normal genetic marker and also a methylation marker (Colot et al. 1996). This experiment was a tour de force, and it showed that the absence of methylation at a given site could be repaired at meiosis: The heterozygous methylation site could become a methylation homozygote.

Jablonka and Lamb (1995) review the evidence for the transmission of epigenetic information throughthegermline,followingmy earlier discussion. There are many examples where the rules of normal genetic transmission and segregation are not evident, but how the epigenetic information is actually transmitted and processed is in most cases unknown. There are many human phenotypes with a strong familial association but which are not inherited in a normal Mendelian fashion. Such traits may be labelled "multifactorial, with variable penetrance", which means that we simply do not understand how the condition is inherited.

There is now accumulating evidence that ionising radiation can induce transgenerational effects, and that at least some of these can be due to heritable changes in DNA methylation (Dubrova et al. 2000; Dubrova 2003; Barber et al. 2000; Morgan 2003; Pogribny et al. 2004). There may well be other environmental influences that have similar effects. Teratogens such as thalidomide interfere with development at a sensitive stage, but it is not known how they act. They presumably target normal cells, or possibly signalling between cells. If teratogens also affect germ line cells, for example by altering DNA methylation, then it is conceivable that their effect could be transmitted to the following generation (Holliday 1998). Although remarkably little is known about the epigenetic system, it is possible to make some comparisons between it and the classical genetic system, and these are summarised in Table 2.

## The Lamarckian Dimension

This theme of transgenerational effects is explored more fully in Jablonka and Lamb's book *Epigenetics and Evolution,* the subtitle of which is *The Lamarckian Dimension* (1995). They argue that epigenetic information may be influenced by the environment, and that some of such changes could, in principle, be transmitted to the next generation. This could have a role in evolution, and thus resurrect the discredited doctrines of Lamarck. This was the challenge that Maynard Smith took up. The models he considers are situations where an organism responds to an environmental stimulus (the thickening of the skin on the human foot is a good example), and that this at some point becomes an inherited trait. He explains that this is exactly what Waddington (1961) meant by "genetic assimilation". Maynard Smith (1989) remarks: "Such a process is interesting, but I do not think it alters greatly our view of evolution." He also points out that the experiments that Waddington carried out to demonstrate genetic assimilation used a non-adaptive trait (cross-veinlessness in response to heat shock). However, he also obtained evidence for genetic assimilation using the adaptive trait of salt tolerance. Jablonka and Lamb discuss a number of possible examples of Lamarckian inheritance, but I think it would be fair to conclude that much more rigorous evidence is required.

During normal development there maybe better examples of stimuli that change the epigenotype of cells. Suppose there is a receptor on a cell surface that senses the presence of a morphogen, hormone or growth factor. The interaction is followed by signal transduction to the nucleus, and a given sequence of DNA is modified by DNA methylation. This in turn changes the expression of an adjacent gene, so that the phenotype of the recipient cell is changed. If this cell also divides with the same epigenotype, then we have a scenario in which one cell, or a group of cells, alters the epigenotype and phenotype of another cell, or group, and this new trait is heritable through mitosis. This is illustrated in Fig. 1. The closest documented example is the synthesis of vitellogenin in response to the hormone estradiol (Jost and Saluz 1993). This is accompanied by the specific loss of methylation in the gluco-corticoid receptor sequence adjacent to the vitellogenin gene. In this case, the cells with the newly induced gene do not divide.

## THE CENTRAL DOGMA OF MOLECULAR BIOLOGY REVISITED

As a working hypothesis in the early years of molecular biology, Crick (1958) proposed the "central dogma". This was that information can flow from nucleic acid to nucleic acid, and nucleic acid to protein, but not from protein to protein, or from protein to nucleic acid. By "information" he meant the sequences of nucleotide bases in nucleic acids and of amino acids in proteins. The central dogma has held up remarkably well, and was not contravened, as some seemed to think, by the discovery of reverse transcriptase. In terms of sequence, the only known exceptions, as noted above, are in the formation and mutation of antibody genes. However, if the definition of information is changed, then the dogma breaks down. Cytosine methylation is an important signal,so the same primaryDNA sequences can haveverydifferent properties. Thus the targeting of DNA methylases to specific sequences, or the removal of methylation, became examples of proteins imposing information on DNA, or removing it. Nowadays, everyone can agree that a continual interaction between nucleic acids and proteins occurs in all organisms, and some of these interactions result in changes of information, in the broadest sense. It is probable that these are most important during the development of multicellular organisms. They are the changes that produce the spectrum of epigenotypes in the complete organism. Once the epigenotype is established, whether in dividing or non-dividing cells, there may be no more interactions between proteins and DNA other than the normal regulation of cellular activity, which includes, of course, the control of the cell cycle.

In tumourigenesis the normal interactions between DNA and proteins break down during a series of sequential steps. This breakdown includes aberrant DNA methylation, and especially the de novo methylation and silencing of tumour suppressor genes (Holliday and Ho 1998; Jones and Baylin 2002). In terms of documented observations, more is now known about these abnormal changes than the normal ones that occur during development and differentiation. This is a challenge for the future: What exactly are the epi-genetic controls that determine the phenotype of any specialised cell? What determines the synthesis of luxury proteins in any given specialised cell, and the repression of all those luxury proteins in all other specialised cells? One approach that is well underway is to determine the exact nature of the epigenome in different cell types (Novick et al. 2002; Beck and Olek 2003).

## Conclusions

It should be accepted that there is dual inheritance in multicellular organisms. One is the classical genetic system, which we now know is based entirely on changes in DNA base sequence. The other is less obvious but nevertheless a very real epigenetic system: It comprises the heritable events in somatic cells that give rise to any specific organism. It also includes the fact that many specialised cells maintain their phenotype though multiple cell divisions. These epigenetic controls comprise information which is superimposed on DNA sequences. Whereas genetics is thoroughly documented and understood, the study of epigenetics is in its infancy. The rules for the unfolding of the programme of development are not understood. Nor are the controls that determine which set of luxury proteins are synthesised in a differentiated cell, and how the genes for the luxury proteins of other cell types are silenced. Much progress has been made in some areas of epigenetics—for example, the mechanism and role of imprinting—but achallenge ofthe twenty-firstcentury is to reveal the nature of many other fundamental epigenetic events. Another type of dual inheritance is better documented. This is the fact that in cultured mammalian cells inheritance cannot only be due to standard gene mutations, but also to heritable changes in DNA methylation. These alternative types of inheritance can occur in derivatives of the very same cell type, such as CHO cells (Paulin et al. 1998). One field of epigenetics that has recently received an extraordinary amount of attention is the study of the de novo methylation and gene silencing in tumour suppressor genes during oncogenesis (Jones and Baylin 2002; Millar et al. 2003). Hopefully this will stimulate equally important studies of DNA methylation in normal cells. Finally, the sequencing of all five bases in the DNA of specialised cells—the epigenome project—will yield a vast amount of important information (Novick et al. 2002). It is likely that there are common methylated sites (e.g. silenced transposable elements) and also the much more interesting specific sites, which will be different in distinct cell types and tissues.

# 10

# Mutagenesis at Methylated CpG Sequences

## INTRODUCTION

DNA methyltransferases. Most or all of these 5-methylcytosine bases are found in the dinucleotide sequence CpG (Riggs and Jones 1983). As discussed in this chapter and elsewhere in this book, CpG methylation may play a critical role in carcinogenesis. Genome-wide decreases and sequence-selective increases in DNA methylation have been found in the DNA of tumor cells, and these changes have been implicated in tumor development (Jones and Baylin 2002). The establishment and maintenance of DNA methylation patterns and the disruption of these patterns in tumors are epigenetic events. On the other hand, the hypermutability of CpG sequences, largely attributed to deamination of 5-methylcytosine, has been considered as one possible source ofgenetic mutation in tumors (Jones 1996; Jones et al. 1992; LairdandJaenisch 1996; Pfeifer 2000).

Historically, 5-methylcytosine was first identified as a spontaneous muta-tional hotspot in *Escherichia coli* more than 25 years ago (Coulondre et al. 1978; Duncan and Miller 1980). Many studies have since confirmed the importance of methylated cytosines as mutational targets. CpG sequences are preferentially mutated in many different human genetic diseases, for instance in the factor IX gene in hemophilia (Krawczak et al. 1998; Sommer 1995). It can be assumed that most of these sequences are methylated in the germ line, although an exact determination of methylation patterns in coding sequences of the mutated genes has rarely been made. In the *HPRT* gene, the most frequent mutational events in dividing somatic cells and in germ cells are C to T substitutions at CpGs (O'Neill and Finette 1998).

These transitions are thought to result from deamination of 5-methylcytosine so that the methylated CpG dinucleotide is viewed as inherently mutagenic. DNA methylation-mediated mutagenic events apparently have had a strong impact on vertebrate genome evolution, since the majority of CpG dinucleotide sequences have been lost. In mammalian genomes, CpGs are present only at about one fifth of their expected random frequency (Schorderet and Gartler 1992; Sved and Bird 1990) so that only about 1% of all DNA bases are 5-methylcytosine. In contrast, a normal frequency of CpGs is maintained at CpG islands—sequences with high G+C content—which probably are not methylated in the germ line and are thus free from trans-generational mutational pressure. In certain tissues, transversion mutations at CpG sequences are characteristically elevated (Knoll et al. 1994; Pfeifer et al. 2002).

In this chapter, we will consider factors thought to be responsible for the high mutation frequencies seen at CpG dinucleotides in mammalian cells.

## The p53 Gene as a Mutation Reporter

Unless one examines the patterns of silent substitutions or pseudogene sequences on an evolutionary scale, most studies of in vivo mutagenesis make use of mutation reporter genes and involve a selectable phenotype. Thus, the analysis will necessarily be constrained by the requirements leading to a selectable phenotype. For some genes, only a few amino acid changes will produce a selectable change and these are less suitable for the analysis of mu-tational spectra. A good mutation reporter system will have a large number of mutational changes that can produce a phenotype. A unique system that fits this category is the p53 gene, which is commonly mutated in human tumors.

Proto-oncogenes and tumor suppressor genes may be critical selectable targets for mutations in cancer cells. The readiness with which CpG sequences in coding sequences undergo mutation will likely be involved in shaping mu-tational spectra in tumors. It has been shown that more than 50% of all human tumors have a mutation in the p53 gene (Greenblatt et al. 1994). This high frequency of mutation provides us with a unique opportunity to investigate the possible origins of these mutations (Greenblatt et al. 1994; Hainaut et al. 2001; Hollstein et al. 1991; Hussain and Harris 1998; Pfeifer et al. 2002). About 300 out of the 393 codons of the p53 gene can harbor mutations according to the p53 mutation database (Olivier et al. 2002). This database currently has close to 20,000 entries and is still growing. Unlike several other tumor suppressor genes, in which nonsense and frameshift mutations predominate, most of the mutations in p53 are missense mutations, thus providing a wider spectrum of mutational events. About 30% of all p53 mutations are found at CpG dinucleotides. CpG sequences in the p53 coding sequence are highly methylated in all human tissues examined (Rideout et al. 1990; Tornaletti and Pfeifer 1995). The majority of p53 mutations are found along its DNA binding domain sequence. There are 23 methylated CpGs, which constitute only about 8% of the central DNA binding domain sequence between codons 120 and 290. However, about 33% of all mutations in this region occur at these relatively few CpG sites. The majority of these p53 alterations are transitions and an even higher percentage of germline mutations (up to 60%) occur at CpG sites in patients with the cancer-prone disease Li-Fraumeni syndrome (Laird and Jaenisch 1996). Therefore, methylated CpG dinucleotides are the single most important mutational targets in p53. Five major p53 mutational hotspots, i.e., codons 175, 245, 248, 273, and 282, all contain methylated CpG dinucleotides.

Human tumors of different tissue origin display different patterns of p53 mutations. In colon cancer, transitions at CpGs account for almost 50% of all point mutations but, strikingly, only 10% of liver or lung cancers contain such mutations. In contrast, in lung and liver cancers, the predominant class of mutations is G to T transversions (Hussain and Harris 1998). Transition mutations at CpG are relatively frequent (generally 20%–25%) in almost all internal cancers except lung and liver. Stomach cancers (33%), brain cancers (38%), and colorectal cancers (46%) have the highest frequencies of CpG tran-sitionmutations according to the InternationalAgencyforResearchonCancer (IARC) p53 mutation database (Olivier et al. 2002). The reason for this tissue specificity of p53 mutagenesis is unknown. The CpG transition mutations have been linked to elevated deamination of endogenous 5-methylcytosine bases (Gonzalgo and Jones 1997; Jones 1996; Jones and Baylin 2002; Laird and Jaenisch 1996).

In skin cancers, transition mutations are largely confined to dipyrimidine sequences. The differences in mutational profiles for different tumor types suggest that exogenous carcinogens are implicated in p53 mutagenesis at least in some tissues. Solar UV light is involved in the induction of nonmelanoma skin tumors, basal cell and squamous cell carcinoma and also melanoma. p53 mutations in these human skin cancers bear C to T and CC to TT transition signatures (Brash et al. 1991; Dumaz et al. 1993; Ziegler et al. 1993), two types of base substitutions specifically induced by UV light in experimental systems (Pfeifer 1997). Benzo(*a*)pyrene, which preferentially damages guanine bases and is an important mutagenic component of tobacco smoke, induces predominantly G to T transversions in murine tumors (Ruggeri et al. 1993). The percentage of G to T transversions in p53 is unusually high in human lung tumors diagnosed in smokers (Greenblatt et al. 1994; Hernandez-Boussard and Hainaut 1998; Pfeifer et al. 2002). Another example links hepatocellular carcinomas from certain areas of the world to a specific action of aflatoxin B1 on the p53 gene (Aguilar et al. 1993; Puisieux et al. 1991).

Interestingly, mutations in lung cancer, but not in hepatocellular carcinoma, also cluster at CpG dinucleotides, although transitions at such sites are only 10% of all mutations. The high transition mutation rate at methylated CpGs in many cancers has been explained by the elevated susceptibility of these sites to spontaneous deamination (see the following section) although other mechanisms are also conceivable. However, it is more difficult to find a sound explanation for the prevalence of transversions at methylated CpGs in carcinogen-induced tumors like lung cancer, if one considers only endogenous sources of mutations in the form of 5-methylcytosine deamination. Interestingly, base changes characteristic for skin cancer, i.e., transitions at CC or TC dipyrimidine sequences, also show an association with methylated CpGs (Tommasi et al. 1997). In later parts of this chapter, we will summarize alternative explanations for the origin of CpG-associated mutations in these human tumors.

## Deamination of 5-Methylcytosine

Deamination of 5-methylcytosine is viewed as the main source of the elevated rate of transitions at CpG sequences (Gonzalgo and Jones 1997; Fig. 1). Both cytosine and 5-methylcytosine can undergo hydrolytic deamination resulting in uracil and thymine, respectively. Hydrolytic deamination occurs at cyto-sine in double-stranded DNA at a very slow rate with a half-life of about 30,000 years at 37 °C and pH 7.4 (Frederico et al. 1990; Lindahl 1993; Shen et al. 1994). The chemistry of cytosine deamination involves hydroxyl ion attack on the cytosine base protonated at the N3 position (Frederico et al. 1993). Deamination of cytosine can be enhanced under acidic conditions and by using chemicals such as sodium bisulfite (Frederico et al. 1990; Wang et al. 1980). 5-Methylcytosine is resistant to bisulfite-induced deamination due to sterical reasons. However, methylation at the 5 position of the base ring facilitates spontaneous hydrolytic deamination to a moderate extent (Ehrlich et al. 1986, 1990; Lindahl 1993; Shen et al. 1994; Wang et al. 1982).

As a result, 5-methylcytosines are deaminated two to four times more rapidly than cytosines (Ehrlich et al. 1990; Shen et al. 1994). For double-stranded DNA the difference was determined to be 2.2-fold (Shen et al. 1994). This twofold enhancement is not sufficient to account for the elevated mutation rates seen at mCpGs. The mutational outcome may be affected by differences in repair of the resulting two base–base mismatches. There may be relatively inefficient repair of T/G mismatches vs U/G mismatches (Neddermann et al. 1996; Schmutte et al. 1995). Uracil in DNA is recognized and excised efficiently by the ubiquitous uracil-DNA glycosylase enzymes. Mammalian cells contain four known uracil DNA glycosylases. The UNG protein is highly conserved and is present in most living organisms (Krokan et al. 2002; Olsen et al. 1989). The other mammalian uracil DNA glycosylases are single-strand selective monofunctional uracil DNA glycosylase (SMUG)1, methyl-CpG binding domain protein (MBD)4, and thymine DNA glycosylase (TDG) (Haushalter et al. 1999; Hendrich et al. 1999; Neddermann et al. 1996). UNG and SMUG1 prefer single-stranded DNA but also act on substrates that contain uracil in double-stranded DNA. TDG and MBD4 are specific for excision of uracil from double-stranded DNA and also remove other bases such as thymines from T/G mismatches, and some damaged pyrimidine bases (Abu and Waters 2003; Boorstein et al. 2001; Hang et al. 1998; Hardeland et al. 2003; Hendrich et al. 1999; Neddermann et al. 1996; Saparbaev and Laval 1998; Waters and Swann 1998; Yoon et al. 2003). While UNG is thought to be primarily responsible for correcting dUMP incorporation events during DNA replication (Kavli et al. 2002; Nilsen et al. 2000), the other three DNA glycosylases may counteract the mutagenic consequences of deamination of cytosine or 5-methylcytosine (Hendrich et al. 1999; Neddermann et al. 1996; Nilsen et al. 2001) and may also repair oxidized and adducted pyrim-idines.

Several proteins have the capacity, at least in vitro, to excise T from T/G mispairs. A mismatch-specific thymine DNA glycosylase identified initially by Jiricny and co-workers (Neddermann et al. 1996) was recently shown to have a broader substrate specificity and removes also etheno-cytosine residues and thymine glycols from DNA (Abu and Waters 2003; Hang et al. 1998; Saparbaev and Laval 1998; Yoon et al. 2003). Mammalian proteins binding to methylated CpG sites have been identified. These proteins contain a conserved MBD domain. One of these methyl-CpG-binding proteins, MBD4 has a T/G mispair-specific DNA glycosylase activity (Hendrich et al. 1999). MBD4 efficiently recognizes and removes thymine from a T/G mispair and excises uracil from a U/G mispair at unmethylated CpG sequences. It is interesting that the function of MBD4 is quite similar to that of TDG, despite a complete lack of sequence homology of the two proteins. When MBD4 was deleted in the mouse, there was a two- to threefold increase in CpG transition mutations in mutational reporter genes (Millar et al. 2002; Wong et al. 2002). A mouse knockout model of TDG has not yet been reported, presumably because of embryonic lethality.

It is currently unclear if these two enzymes are the only activities that operate on T/G mismatches derived from deamination of 5-methylcytosines in vivo. Some of these mismatches may be corrected by the general mismatch repair system as well (Bill et al. 1998), although it is unclear how strand specificity of the repair reaction can be achieved. Mammalian homologs of the bacterial very short patch repair (*vsr*) gene product, which corrects T/G mismatches arising at *dcm* methylation sites through an endonucleolytic activity (Hennecke et al. 1991; Lieb 1991; Sohail et al. 1990), have not yet been identified. Since the T/G mismatch is probably repaired less efficiently than a U/G mismatch, this consequently may create a higher risk for mutation fixation.

On the other hand, the rate of CpG germ-line mutation in primate species was estimated to be about 1,250 times slower in an Alu element in p53 intron 6 than the in vitro deamination rate of 5-methylcytosine in double-stranded DNA (Yang et al. 1996b). The germ-line mutation rate was calculated to be even slower at CpGs in the factor IX gene (Sommer 1995). This implies that the existing cellular repair mechanisms may correct not only U/G but also T/G mismatches quite efficiently or that deamination of 5-methylcytosine in vivo is much slower than deamination in vitro.

In fact, it is not proven beyond doubt that the spontaneous deamination model accurately reflects all mutagenesis events at CpG sequences in mammalian cells. One major dilemma is exemplified by the calculation that only two 5-methylcytosines may deaminate per day in eachcell (Schmutte andJones 1998). These numbers appear almost insignificant compared to steady-state levels that have been measured for many endogenous and exogenous DNA lesions, which can be between hundreds and several thousands per cell (Holmquist 1998; Marnett and Burcham 1993).

It is possible that certain chemicals may promote 5-methylcytosine deam-ination at CpGs. Nitric oxide was shown to increase the rate of G/C to A/T transitions in *Salmonella* perhaps via stimulation of deamination (Wink et al. 1991). Direct assays have yet failed to show any significant deamination of 5-methylcytosine by nitric oxide (Felley-Bosco et al. 1995; Schmutte et al. 1994). On the other hand, there is a dose-response relationship between the frequency of G/C to A/T transitions at CpGs in the p53 gene and increased nitric oxide synthase (NOS)2 expression in human colon carcinomas (Ambs et al. 1999) and transition mutations at codon 248 of the p53 tumor suppressor gene could be induced by a nitric oxide-releasing compound (Souici et al. 2000). There are examples of other mechanisms that may affect cy-tosine deamination directly or via formation of intermediates. For example, 5-methylcytosine can be deaminated by a photo-chemical process (Privat and Sowers 1996).

Oxidative damage to 5-methylcytosine results primarily in formation of the deaminated product thymine glycol through a 5-methylcytosine glycol intermediate. Thymine glycol is primarily a replication-blocking lesion (Zuo et al. 1995). However, thymine glycol can be bypassed by DNA damage-tolerant polymerases such as DNA polymerase n_, £,, and κ with incorporation of adenine opposite the lesion (Fischhaber et al. 2002; Johnson et al. 2003; Kusumoto et al. 2002). Oxidative damage-induced transition mutations at CpG sequences are enhanced by CpG methylation (Lee et al. 2002). Interestingly, thymine glycol in the context of an mCpG sequence is recognized and excised by both TDG and MBD4 proteins, pointing to a potential role of this pathway in CpG mutagenesis (Yoon et al. 2003). In nucleotide excision repair-deficient cells, oxidative DNA damage produces mCpG to TpT tandem mutations (Lee et al. 2002), which may be generated from a cross-link lesion between 5-methylcytosine and guanine (Zhang and Wang 2003). Oxidation of the 5-methyl group of 5-methylcytosine is also a possibility (Burdzy et al. 2002; Rusmintratip and Sowers 2000) and generates 5-hydroxymethylcytosine and 5-formylcytosine. 5-Hydroxymethylcytosine is not mutagenic and is present as a normal base in some bacteriophages (Wyatt and Cohen 1953). The mutational specificity of 5-formylcytosine is broad, and includes targeted (5-fC^G, 5-fC^A, and 5-fC^T) and untargeted mutations (Kamiya et al. 2002b). Deamination of 5-hydroxymethylcytosine and 5-formylcytosine generates 5-hydroxymethyluracil and 5-formyluracil. These oxidized bases pair primarily with adenine during replication [although 5-formyluracil is more promiscuous; (Kamiya et al. 2002a)] and, as a result, this oxidation-deamination pathway could lead to 5-methylcytosine thymine transitions.

Reactive oxygen species are generated during inflammatory responses by neutrophils and phagocytes, and this could be a risk factor for cancer (Hal-liwell 2002; Jackson and Loeb 2001). Of relevance to a possible involvement of oxidative stress in CpG mutagenesis is the fact that there is a dramatic increase in CpG transition mutations in cancers associated with an inflammatory response, such as *Scfa'stosoma*-associated bladder and rectal cancers, ulcerative colitis associated colon cancers, and esophageal cancers in certain geographic areas (Ambs et al. 1999; Biramijamal et al. 2001; Hussain et al. 2000; Sepehr et al. 2001;

Warren et al. 1995; Zhang et al. 1998). Glyoxal, a known mutagen, has been shown to directly deaminate 5-methylcytosine to thymine at a higher rate than it deaminates cytosine to uracil(Kasai et al. 1998).

It has also been reported that ethylene oxide, a rodent and probable human carcinogen, and 1-nitropyrene, an environmental muta-gen, have a capacity to promote cytosine deamination (Li et al. 1992; Malia and Basu 1994). Compounds that intercalate into the DNA double helix at methylated CpG sites mayhave the capacityto promote deamination bycreatingpar-tially unwound stretches of DNA. Effects on deamination of 5-methylcytosine have not been measured for most of these compounds. An additional possibility that warrants consideration is that nuclear proteins binding at or near mCpG sequences may enhance deamination of 5-methylcytosine.

## Enzymatic Deamination Reactions

An alternative pathway may involve the intrinsic mutagenic capacity of the enzymatic de novo methylation reaction at CpG sequences. Using in vitro systems, it has been demonstrated that several bacterial methyltransferases, including *Hpa*II, *Sss*I, and others, promote C to U deaminations at CpG targets at low concentrations of the methyl group donor *S*-adenosyl-l-methionine (Shen et al. 1992; Wyszynski et al. 1994; Yang et al. 1995). The methyl group transfer catalyzed by the *Hha*I methyltransferase was shown to occur through formation of an active intermediate between a cysteine residue of the enzyme and position 6 of a cytosine base swung completely out of the DNA helix (Klimasauskas et al. 1994). The half-life of this intermediate may increase when the concentration of *S*-adenosylmethionine is low. This and the demonstrated higher affinity of DNA methyltransferase towards T/G and U/G mismatches than towards normal C/G base pairs (Gonzalgo and Jones 1997; Klimasauskas et al. 1994; Yang et al. 1995) together may provide an enzyme-mediated mechanism leading to the hypermutability of CpG dinucleotides. One bacterial methyltransferase was shown to convert 5-methylcytosine directly to thymine (Ye bra and Bhagwat 1995). The proposal that enzyme-catalyzed events may play a role in carcinogenesis is supported by a number of studies reporting elevated expression of cytosine DNA methyltransferase in human colon cancer cell lines and in colonic mucosa (El-Deiry et al. 1991; Schmutte et al. 1996). It is not clear, however, whether enzyme-mediated deamination is a significant event in vivo, where the concentration of methyl-group donors is high (Wyszynski et al. 1994). The extent of the involvement of enzyme-mediated deamination in CpG mutagenesis requires additional investigation.

Another more direct pathway to 5-methylcytosine deamination may involve cytosine deaminases. Activation-induced cytidine deaminase (AID) is required for somatic hypermutation of immunoglobulin genes (Muramatsu et al. 2000). Although AID has sequence similarity to an RNA-editing enzyme, APOBEC-1, it is unknown how AID is precisely functioning in somatic hy-permutation. Expression of AID in *E. coli* produces nucleotide transitions at dC:dG base pairs (Petersen-Mahrt et al. 2002). Mutation triggered by AID is enhanced by a deficiency of uracil-DNA glycosylase, which suggests that AID functions by deaminating dC residues in DNA (Di Noia and Neuberger 2002). Similarly, APOBEC1 and its homologs APOBEC3C and APOBEC3G exhibit potent DNA mutator activity in the *E. coli* assay. Each protein has a certain target sequence specificity (Harris et al. 2002). The AID-induced deamination reaction seems to favor single-stranded DNA, as it occurs, for example, during the process of transcription (Pham et al. 2003; Ramiro et al. 2003; Sohail et al. 2003). If mis-targeted in the genome, the cytidine deaminases could produce mutations leading to cancer. In fact, some AID family members are expressed preferentially in cancer tissue (Harris et al. 2002), and APOBEC1 transgenic mice have an increased incidence of hyperplasia and liver cancer (Yamanaka et al. 1995). AID and APOBEC homologs deaminate not only cytosine but also 5-methylcytosine in vitro andin*E. coli* (Morgan et al. 2004). However, recently it has been reported that methylation protects cytosines from deamination (Larijani et al. 2005)

## METHYLATED CPG SEQUENCES AS PREFERRED TARGETS FOR MUTAGENS AND CARCINOGENS

It is widely accepted that mutagenesis induced by endogenous and exogenous agents is an important component of tumorigenesis. This concept was well proved experimentally in animal models in which mutagen exposure led to tumors harboring carcinogen-specific ("fingerprint-type") mutations in *ras* genes or in the p53 gene (Barbacid 1987; Ruggeri et al. 1993). For reproducing the p53 mutational spectrum that occurs in human cancers, we are necessarily left with more indirect approaches. One approach involves identification of sequence-specific DNA lesions generated by carcinogens in the p53 gene, and correlation of these "fingerprints" with p53 mutations collected from human cancer databases (Pfeifer et al. 2002). This approach is based on mapping of DNA damage at nucleotide resolution by the ligation-mediated PCR (LMPCR) technique (Denissenko et al. 1996; Pfeifer et al. 1991). Using this technique, we have compared the distribution of DNA damage in the p53 gene of human cells exposed to UV light, benzo(*a*)pyrene diolepoxide (BPDE), or aflatoxin B1 (AFB1) with the distribution of p53 mutations in human cancers of the skin (non-melanoma), lung, and liver (Denissenko et al. 1998a, 1996; Pfeifer et al. 1991; Tommasi et al. 1997; Tornaletti and Pfeifer 1994). These experiments revealed a previously unrecognized role of methylated CpG sites as preferential targets for physical and chemical genotoxic agents.

Exposure to solar radiation is a principal factor in the development of skin cancer (Mortimer 1991). Mutations in the p53 gene were found in a large fraction ofhumannon-melanoma skin tumors and in precursor lesions(Brash et al. 1991; Dumaz et al. 1993; Ziegler et al. 1994; Ziegler et al. 1993). Even normal sun-exposed skin contains a large number of clonal patches of p53-mutated keratinocytes (Jonason et al. 1996). The vast majority of base changes in skin lesions are C to T or CC to TT mutations at dipyrimidine sequences.

These mutations are consistent with the specificity of the most mutagenic UV-induced lesion in mammalian cells, the cyclobutane pyrimidine dimer (Pfeifer 1997). p53 mutations in skin cancer are clustered at several mutational hotspots (Tommasi et al. 1997). With the exception of codons 177 and 278, all the other skin cancer mutation hotspots (codons 152,196, 213, 245, 248, and 282) contain the mutated dipyrimidine in the sequence context 5'CmCG or 5'TmCG. Base changes characteristic for skin cancer, i.e., transitions at CC or TC dipyrimidine sequences, show a strong association with methylated CpGs (Tommasi et al. 1997; You et al. 2001). The relative contribution of p53 mutations affecting dipyrimidines within mCpG sequences is 130/362 (36% of the total mutations), despite the fact that 5'CCG and 5'TCG occur only 19 times in the 1,000-bp double-stranded target area between codons 120 and 290. Importantly, all these CpG sequences are methylated in human keratinocytes (Tornaletti and Pfeifer 1995).

Using 254-nm UVC light for irradiation, we initially found that only some of these skin cancer hotspots were highly susceptible to UV damage formation (Tornaletti et al. 1993). In particular, a lack of correlation was noted at dipyrimidine sequences that contained 5-methylcytosine. Subsequently, we found that mutation hotspot positions that contain 5-methylcytosine within dipyrimidine sequences are up to 15-fold more susceptible to pyrimidine dimer formation after exposure to natural sunlight (Tommasi et al. 1997). Another study has reported a similar phenomenon in human cells irradiated with UVB (280-320 nm), a component of natural sunlight that reaches the earth's surface (Drouin and Therrien 1997). Methylation of cytosine enhances pyrimidine dimer formation by sunlight by 5- to 15-fold (Tommasi et al. 1997). This difference may be explained by the higher energy absorption by 5-methylcytosine compared to cytosine in DNA. The Amax of 5-methylcytosine vs cytosine is red-shifted by about 6 nm so that the Amax of 5-methylcytosine is 273 nm compared to 267 nm for cytosine at neutral pH. This red-shift results in a wavelength-dependent 5-15 times higher extinction coefficient for 5-methylcytosine vs cytosine at wavelengths from 300 nm to 315 nm. This critical part of the solar spectrum is a component of sunlight that reaches the earth's surface and is absorbed by DNA. In addition, 5-methyl-thymidine monophosphate (dCMP) has a significantly lower excited singlet state energy than TMP or deoxycytidine monophosphate (dCMP) (Ruzcicska and Lemaire 1995) and therefore 5-methyl-dCMP may be a singlet energy trap in DNA. The fact that sunlight induces cyclobutane pyrimidine dimers preferentially at 5-methylcytosine bases was not recognized previously but could have important implications for sunlight-induced mutagenesis, not limited to the p53 gene. Consistent with the enhanced formation of pyrimidine dimers at 5-methylcytosines, simulated sunlight induces mutational hotspots at dipyrimidine sequences containing 5mC (You et al. 2001; You et al. 1999). Moreover, the 5-methylcytosine bases within pyrimidine dimers are prone to hydrolytic deamination, which increases their mutagenicity (Lee and Pfeifer 2003; Tu et al. 1998). These deamination reactions maybe followedbyacorrect polymerase bypass of the deaminated dimers during DNA replication with incorporation of deoxy-ATP (dATP) and targeted C or 5mC to T transition mutations, the predominant types of mutation seen in nonmelanoma skin tumors.

Tobacco smoking is a strong risk factor for the development of lung cancer (Hecht 1999). The signature of p53 lung tumor mutations consists of G to T transversions biased to guanines on the nontranscribed DNA strand (Hussain and Harris 1998; Pfeifer et al. 2002). G to T transversions are typical for bulky adduct-forming mutagens including the class of polycyclic aromatic hydrocarbons (PAHs). Benzo[*a*]pyrene is a widely studied member of the PAH class. Upon metabolic activation to BPDE, it induces G to T mutations (Chen et al. 1990). The distribution of BPDE adducts along the p53 gene was mapped at nucleotide resolution in carcinogen-treated normal human bronchial epithelial cells (Denissenko et al. 1996). Selective adduct formation sites were major mutational hotspots in human lung cancers, i.e., there was an excellent correlation between the benzo[*a*]pyrene adduct spectrum and the mutation spectrum in lung cancer (Pfeifer et al. 2002). We have shown that the mechanistic basis for the selective occurrence of these damage hotspots is related to patterns of cytosine methylation in the p53 gene (Denissenko et al. 1997). The distribution of BPDE-DNA adducts differed drastically in CpG-methylated DNA compared to non-methylated DNA. Guanines 3' to 5-methylcytosines were the preferentially adducted positions, and CpG methylation strongly enhances BPDE adduct formation (Chen et al. 1998; Denissenko et al. 1997; Tretyakova et al. 2002; Weisenberger and Romano 1999). Therefore, CpG din-ucleotides, which are methylated in the human p53 gene in all human tissues examined (Tornaletti and Pfeifer 1995), in addition to being an endogenous promutagenic factor, represent preferential targets for exogenous chemical carcinogens as well. Figure 1 shows possible pathways operating at methylated CpG sites and creating increased mutation rates.

The ability of a PAH diolepoxide to form intercalative non-covalent complexes with DNA prior to covalent binding should be an important factor in determining the reactivity of these compounds. In the case of BPDE, hy-drophobic effects (Geacintovetal. 1988) or increased molecular polarizability and base stacking (Sowers et al. 1987) derived from the methyl group of 5-methylcytosine seem to facilitate the creation of an intercalation site. The increase in BPDE intercalative binding to methylated CpG sites is then eventually reflected in the extent of covalent interactions. Likewise, other DNA adducts that arise through intercalation may form more easily at methylated CpGs (Chen et al. 1998; Parker et al. 2004).

The extent by which enhanced binding of an individual carcinogen at methylated CpGs affects mutagenesis at the same location has been studied in mouse cells carrying the *lacI* and *cII* transgenes. These cells were treated with BPDE and the mutations were scored. A dominant fraction of the mutations (58%-77% of all G to T mutations) occurred at methylated CpG sequences (Yoon et al. 2001). The G to T transversion mutation hotspots observed in this system were strikingly similar to the ones observed in lung tumors from cigarette smokers.

The palindromic structure of a methylated CpG site creates two possible sources of a C to T mutational event. The observed mutations maybe caused by a lesion residing preferentially at guanine bases in methylated CpG sequences, which would produce G to A transition mutations indistinguishable from C to T mutations on the opposite strand. However, compounds with this mutational specificity have not yet been identified. Certain mutagens may preferentially modify or form adducts at methylated cytosines themselves and cause transition mutations by mispairing or may increase the hydrolytic deamination of 5-methylcytosine, although such mutagens have not yet been specifically identified either. These possible mechanisms are outlined in Fig. 1.

Another crucial component of the mutagenesis process is DNA repair. The presence of 5-methylcytosine may affect the sequence-dependent repair of DNA adducts. A 5-methylcytosine base at the 5'-position adjacent to an O6-methylguanine lesion strongly diminished repair of this lesion by O6-methylguanine DNA methyltransferase (Bentivegna and Bresnick 1994). On the other hand, the presence of 5-methylcytosine protects neighboring gua-nines from O6-methylation (Ziegel et al. 2004). Indeed, methylating agents that produce O6-methylguanine induce transition mutations preferentially at guanines that are *not* part of a methylated CpG sequence (Bodell et al. 2003) suggesting that inhibition of repair does not play a major role in this phenomenon.

Most pyrimidine dimers, which contain a 5-methylcytosine as the dimerized base, were repaired more slowly than neighboring sequences without 5-methylcytosine, and this correlated with the position of p53 mutation hotspots in skin tumors (Tornaletti and Pfeifer 1994). Further work is required to investigate more directly the influence of methylation on repair of DNA lesions at CpG sites by the base excision repair, mismatch repair, and nucleotide excision repair complexes.

# 11

# Post-replicative DNA Modification

Cytosine methylation is a common form of post-replicative DNA modification seen in both bacteria and eukaryotes. Modified cytosines have long been known to act as hotspots for mutations due to the high rate of spontaneous deamination ofthis base to thymine, resulting in a G/T mismatch. This will be fixed as a C'!T transition after replication if not repaired by the base excision repair (BER) pathway or specific repair enzymes dedicated to this purpose. This hypermutability has led to depletion of the target dinucleotide CpG outside of special CpG islands in mammals, which are normally unmethylated. We review the importance of C'!T transitions at non-island CpGs in human disease: When these occur in the germline, they are a common cause of inherited diseases such as epidermolysis bullosa and mucopolysaccharidosis, while in the soma they are frequently found in the genes for tumor suppressors such as p53 and the retinoblastoma protein, causing cancer.

We also examine the specific repair enzymes involved, namely the endonuclease Vsr in *Escherichia coli* and two members of the uracil DNA glycosylase (UDG) superfamily in mammals, TDG and MBD4. Repair brings its own problems, since it will require remethylation of the replacement cytosine, presumably coupling repair to methylation by either the maintenance methylase Dnmt1 or a de novo enzyme such as Dnmt3a.

Uncoupling of methylation from repair may be one way to remove methylation from DNA. We also look at the possible role of specific cytosine deaminases such as Aid and Apobec in accelerating deamination of methylcytosine and consequent DNA demethylation.

## INTRODUCTION

### Intrinsic Instability of Methylcytosines Due to Deamination

Loss of the amino group or deamination occurs spontaneously for several of the nucleotide bases that make up DNA. The rate of deamination is highest for cytosine of the four standard nucleotides and is estimated to occur in one of every $10^7$ cytosine residues per day (Ehrlich et al. 1986). The product of this reaction is uracil (Fig. 1), which can base pair with adenine and direct incorporation of the latter following replication, thus leading to a C to T transition. However, uracil is not normally found in DNA and so can easily be recognized and removed by repair systems in the cell.

If uracil were a normal component of DNA, then recognizing the products of cytosine deamination would be more difficult, and a gradual loss of cytosines from the DNA would be expected over evolutionary time. This is thought to be the main reason for the use of thymine in DNA, rather than the uracil found in RNA. If theories regarding the RNA origins of life are correct, then the adoption of thymine rather than uracil in DNA would represent a major stepping stone in the development of a system for long-term stable storage of genetic information. The danger of having a naturally occurring base generated by spontaneous deamination is illustrated in the case of 5'-methylcytosine (5meC). Methylation of cytosine is a common post-replicative modification of DNA in both bacteria and eukaryotes and occurs at the 5' position of the pyrimidine ring. Deamination of 5meC generates thymine, which is normally found in DNA and so would be harder to recognize.

Spontaneous deamination of 5meC occurs at an approximately fivefold higher rate than for native cytosine (Ehrlich et al. 1986) and an estimated four 5meC residues deaminate per diploid cell per day (Shen et al. 1994). The thymine will cause the incorporation of an adenine on the opposite strand at replication, fixing a C to T transition in the DNA. Cytosine methylation is found in bacteria, animals, and higher plants, though the sequence context in which the methylated cytosine resides varies. In *Escherichia coli*, cytosine methylation occurs at the internal cytosines in the palindromic target sequence CCWGG, where the third residue is either an A or a T (May and Hattman 1975).

Methylation of these cytosines is carried out by the bacterial DNA cytosine methyltrans-ferase encoded by the *Dcm* gene. Most bacterial methyltransferases form part of a restriction/modification (RM) system that is a crucial defense against invading viral DNA (Wilson and Murray 1991).

Methylation of the target sequence in the bacterium's DNA prevents cleavagebyrestriction endonucleases that specifically recognize these palindromes. The absence of methylation on invading viral DNA causes its cleavage and subsequent degradation. While Dcm is an orphan methylase, its target sequence is recognized by EcoRII, and loss of Dcm leads to susceptibility to cleavage by the latter enzyme (Schlag-man et al. 1976; Takahashi et al. 2002). This crucial host defense mechanism may explain why cytosine methylation is retained by bacteria in the face of high rates of spontaneous mutation at methylated cytosine residues. Indeed, methylated cytosines form hotspots for transition mutations in both bacteria and eukaryotes (Cooper and Krawczak 1993; Lieb 1991).

In eukaryotes, cytosine methylation can occur at low levels on target sequences such as CpNpG (Clark et al. 1995; Gruenbaum et al. 1981) and CpT (Lyko et al. 2000; Gowher et al. 2000), but the vast majority is foundat CpG din-ucleotides (the p represents the phosphate linkage). This short palindrome ensures that the target site for methylation occurs on both strands of the DNA. Methylation appears to play a variety of roles in eukaryotes, where it not only helps to maintain repression of viral and transposon promoters in a host-defense role reminiscent of that in bacteria (Jahner et al. 1982; Walsh et al. 1998; Bourc'his and Bestor 2004), but it is also involved in silencing of endogenous genes such as those subject to imprinting (Li et al. 1993) or X-inactivation (Beard et al. 1995).

It also appears to be crucial for stability of pericentric repeats in eukaryotes (Xu et al. 1999) and is involved in other silencing phenomena in fungi, such as methylation induced premeiotically (MIP) (Malagnac et al. 1997) and repeat-induced point mutation (RIP) (Fre-itag et al. 2002). Although there has been some debate over which may be the "primary" function of cytosine methylation in eukaryotes, especially vertebrates (Bird 1997; Yoder et al. 1997), there is no doubt that it plays a crucial role in the healthy organism and that strong selective pressure operates to maintain cytosine methylation in eukaryotes, since mutations in the DNA methyltransferases are typically lethal for the organism (Li et al. 1992; Okano et al. 1999).

## DISCOVERY OF G/T MISMATCH-SPECIFIC REPAIR IN BACTERIA

Deamination of cytosine leads to the presence of uracil in the DNA, which is of course not normally found there. This aberrant base is recognized by an enzyme from the uracil DNA glycosylase (UDG) family, w hich forms part of the base-excision repair pathway (Pearl 2000). These enzymes recognize lesions in the DNA such as the products of deamination and cut out the aberrant base by cleavage of the *N*-glycosyl bond attaching the base moiety to the deoxyribose-phosphate backbone. Uracil glycosylases can recognize the U/G mismatch and selectively remove the uracil base, leaving an apyrimidinic or AP site.

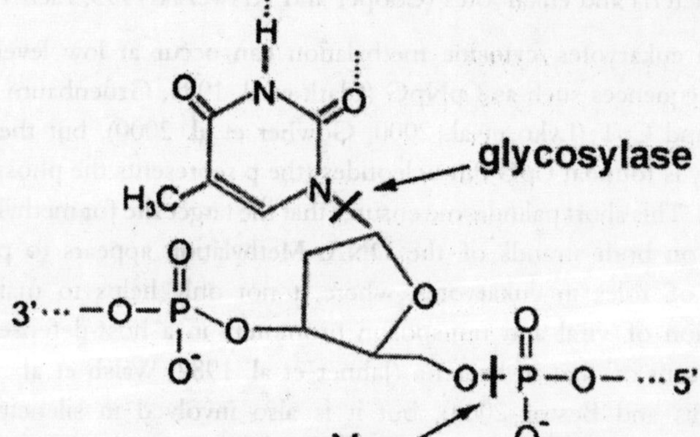


***Fig. G/T mismatch-specific repair enzymes in bacteria and eukaryotes have different cleavage sites. Vsr is an endonuclease, cleaving the phosphodiesterbond and excising the nucleotide, while the glycosylases such as TDG and MBD4 cut the* N*-glycosidic linkage between the base moiety and the phosphodiester backbone, resulting in an apyrimidinic (AP) site***

This is subsequently repaired by removing the deoxyribose 5-phosphate and replacing the whole nucleotide before ligating the backbone together. The presence of an unnatural U/G base pair is a unique signal that allows selective repair of the deaminated cytosine in DNA: Uracils are not removed from RNA molecules or thymines from DNA. Since the uracil is not normally a component of DNA, the guanine is clearly the undamaged base and is used as the template for repair. However, for the T/G mismatch generated by deamination of methylcytosine, the question remains: How does the cell know which is the incorrect base?

Two groups (Shenoy et al. 1987; Zell and Fritz 1987) addressed this problem in bacteria by generating an M13 phage DNA heteroduplex containing a T/G mismatch in the context of the Dcm-methyltransferase recognition site CCWGG. Introduction of hetero duplexes into *E. coli* gave efficient repair of the mismatch with a high bias towards removal of the thymine. This novel enzyme activity was also boosted by the presence of a 5meC on the opposite strand. These experiments revealed the presence of a dedicated repair system termed very short patch (VSP) repair whose key component is a mismatch-specific sequence and strand-specificendonuclease that acts on the T/G mismatches generated by spontaneous deamination of the methylcytosine. This endonuclease "assumes" the T is the incorrect base in the G/T mismatch and removes it.

The gene encoding VSP activity, *vsr,* was discovered close to the *dcm* gene (Sohail et al. 1990). In an elegant set of experiments, Margaret Lieb was able to show that deletion of both genes removed a mutation hotspot at a methylated cytosine in the *lacI* gene: reintroduction of *dcm* without the *vsr* gave a tenfold higher rate of mutation at this site, an effect which could be rescued by the addition of *vsr* (Lieb 1991). The Vsr protein was isolated by Fritz and colleagues (Hennecke et al. 1991) and shown to be a novel endonuclease that generates a single strand break 5' to the thymines produced by methyl cytosine deamination in the Dcm recognition site.

The endonuclease activity is stimulated by the *MutS* and *MutL* genes, whose products act together to form a sensor for mismatched base pairs, and depends on *polA,* the DNA polymerase gene, which will remove the mismatched nucleotide and replace it with a cytosine. The importance of having efficient repair of deaminated methylcytosines is illustrated by the fact that *vsr* is not only tightly linked to the *dcm* gene but actually overlaps it by six codons, and the two proteins are made separately from a single RNA transcript (Dar and Bhagwat 1993). Vsr is an endonuclease that cleaves the phosphodiester backbone of the DNA, rather than excising the pyrimidine base as seen with the UDG family, and appears to have no homologs in eukaryotes. Instead, T/G mismatch repair in higher organisms involves members of the UDG superfamily and may have been acquired as an additional specificity by enzymes whose primary function appears to be U/G repair (Gallinari and Jiricny 1996).

Interestingly, bacteria discriminate between G/T mismatches formed by methylcytosine deamination and those formed by replication errors. This is important, since during replication it may be the G or C that has been misincorporated, while 5meC deamination always mutates the C. While the latter type of error is repaired using VSP, which is part of the base excision repair (BER) pathway, the former are repaired by the more familiar long patch repair system, also known simply as mismatch repair. In *E. coli,* this uses MutL and MutS, but in addition requires MutH, which senses which strand is newly replicated (Bhagwat and Lieb 2002).

MutH does this by determining which chain carries methylated *adenine* sites, so this is sometimes (confusingly) called methylation-directed mismatch repair (MMR), but this is here referring to the use of adenine methylation as a tag for the template strand. This system seems to be confined to some gram-negative bacteria, however, since no MutH homologs have been found in other organisms. Long patch repair, as the name suggests, involves removal of a long stretch of DNA carrying the mismatch using one of a variety of exonucleases.

## Under-Representation of CpG Dinucleotides Caused by 5meC Loss and Rise of CpG Islands in the Mammalian Genome

Vsr is more active during the stationary phase of the bacterial life cycle (Bhag-wat and Lieb 2002) and efficiently repairs T/G mismatches specifically in the Dcm target sequence that arise from deamination of the methylated cyto-sine. Misincorporation of a T opposite a G (or vice versa) during replication is more likely to be repaired using the mismatch repair system, which does so more efficiently during the growth phase.

However, if a T/G mismatch is missed by the latter system during growth, Vsr may act on this mismatch during the stationary phase. Since Vsr has a high preference for removal of the thymine, this system can have detrimental side effects when it is the gua-nine that is incorrect: Sites such as 5'CTAG/3'GGTC, where a G has become misincorporated opposite a T, can also be recognized by the enzyme, which then replaces the correct nucleotide T with a C, causing a T to C transition.

Analysis of tetranucleotide frequency in bacteria showed a relative depletion of CTAG and increase in CCAG sequences consistent with the repair activity altering the sequence composition of the genome (Merkl et al. 1992; Bhagwat and McClelland 1992). It is perhaps in order to reduce this effect that Vsr is produced at lower levels during the growth phase than stationary phase: As a consequence, while it can reduce 5meC to T transitions at Dcm targets by a factor of 4, removal of DNA methylation completely further reduced this type of transition mutation by an order of magnitude (Lieb 1991; Lutsenko and Bhagwat 1999). During stationary phase on the other hand, Vsr may completely prevent mutational hotspots at 5meC (Bhagwat and Lieb 2002). Inefficient repair during the growth phase is thus the main reason for 5meC hypermutability in bacteria.

Cytosine methylation is therefore doubly muta-genic in bacteria: Not only do higher transition rates and inefficient repair lead to increased transitions at methylation target sites, but interference with the MutHLS system also leads to increased fixation rates for some mutations. A similar remodeling of the mammalian genome has occurred as a result of cytosine methylation in eukaryotes, although in this case it appears to be primarily due to inefficient repair, since there is a relative depletion of the methylation target site CpG in the genome as a whole. Given the GC content of the average mammalian genome, CpG sites are present at about 20% of the expected frequency (Sved and Bird 1990). The CpGs are also distributed in a non-random fashion (Bird et al. 1985). Most of the genome is very CpG-poor, with the dinucleotide occurring roughly once every 100 nucleotides, but there also exist short islands of approx. 1 kb containing roughly 10 CpGs per 100 nucleotides.

This represents the expected frequency in these islands, since they also have elevated G:C content relative to non-island DNA (67% vs

41%), so CpGs are being tolerated at the expected level here rather than selected for. As might be expected, these CpG islands are almost always methylation free, particularly in the germ cells (Cross and Bird 1995; Walsh and Bestor 1999) and will therefore not be susceptible to C to T transitions due to deamination of the methylcytosine and not under any greater evolutionary pressure to alter sequence than any other dinucleotide. CpGs outside of the island sequences are almost always methylated, including in the germ line and therefore liable to undergo C'!T transitions at high rates (Cross and Bird 1995; Walsh and Bestor 1999). It is therefore the differential targeting of cytosine methylation in the mammalian genome that has altered the genomic structure here too, though in a more dramatic way than seen in bacteria.

What then are the differences between CpG island and non-island sequences? Islands are almost always associated with the promoters or regulatory regions of genes, so much so that it has become one of the best criteria for identifying promoters. However, not all genes have an associated CpG island, and in fact these are only found at about 60% of human genes, with the remaining 40% being CpG-poor like the rest of the genome (Antequera 2003). Recent surveys of the GC content around the transcriptional start site in different species showed high GC bias in vertebrates, but an AT bias in *Drosophila*, which has almost no methylation (Aerts et al. 2004).

CpG islands may have arisen in the context of this elevated GC content in vertebrates, and primordial islands can be detected near some fish genes. The widespread use of DNA methylation in humans and mice has caused further skewing and the rise of true CpG islands. These islands also show increased nuclease sensitivity, a deficiency in histone H1, hyperacetylation of histones H3 and H4, and nucleosome-free regions, suggesting that they are associated with "open" chromatin, which is easily accessible by *trans*-acting factors (Tazi and Bird 1990; Gilbert and Sharp 1999).

Genes with associated islands tend to have widespread expression (so-called "housekeeping" genes) or to be expressed early in development, and recent genome-wide surveys confirm that there is a good correlation between how widely expressed a gene is and the CpG bias at the transcriptional start (Aerts et al. 2004).

Since methylation of completely unmethylated DNA (de novo methylation) in somatic cells occurs only during early embryogenesis (Okano et al. 1999), a CpG island may mark a gene that needs to establishan accessible chromatin state early in development. The mechanism bywhich CpG islands are maintained in a methylation-free state is currently unclear but is thought to involve binding ofproteins to these regions, blocking methylation. Examples of such factors may include SP1 (MacLeod et al. 1994; Simonsson and Gurdon 2004) and CTCF: Depletion of the latter in the early embryos results in de novo methylation of the CpG island it binds to at the *H19* locus (Fedoriw et al. 2004). Problems exist with this theory, however, since deletion of *Sp1* in the mouse does not result in aberrant methylation of target CpG islands (Marin et al. 1997) and footprinting and nuclease accessibility studies indicate that many islands are more accessible to proteins (and thus presumably DNA methyltransferases), not less (Tazi and Bird 1990).

There are some CpG islands that are exceptions to the methylation-free rule. Prominent examples include CpG islands associated with the control regions of imprinted genes (Bartolomei and Tilghman 1997) and the promoters of genes on the inactive X chromosome (Heard et al. 1997). Here methylation is used to maintain silencing on one allele in the soma, and removal of methy-lation essentially results in reactivation of the inactive copies of imprinted genes (Li et al. 1993; Bourc'his et al. 2001), though the situation is somewhat complicated by antisense control mechanisms used at some imprinted loci. Methylation of CpG islands at the MAGE genes is also seen in somatic tissues, but they are unmethylated and expressed in the male germ line (De Smet et al. 1999). A growing number of genes with CpG islands have also been shown to become inappropriately methylated in specific tumor types (see El-Osta 2004; Bestor 2003; Herman and Baylin 2003 for recent reviews). As we have seen, deamination of methylated cytosines in germ cells leads to C'!T transitions and would tend to erode methylated CpGs over time, as clearly shown in the genomic profile of vertebrates. For imprinted genes, which show methylation in the mature gametes, efficient repair of G/T mismatches to restore the cyto-sine would be needed to allow methylation to be maintained on these islands, but there are no existing studies addressing whether any preferential repair occurs at such facultatively methylated CpG islands. While there is extensive evidence of preferential repair of genes that are being transcribed, this mechanism could not apply to the silent copy of a gene with a methylated CpG island.

Non-CpG island sequences will in general be subject to DNA methylation and subsequently vulnerable to C'!T transitions at a higher rate than seen in CpG islands. This includes the bulk of mammalian DNA, including those genes with non-CpG island promoters. Since CpG islands are normally free of methylation and non-CpG island genes contain low numbers of the target dinucleotide, the main target of DNA methyltransferases is the remainder of the genome, which largely consists of repetitive elements of various types (Yo der et al. 1997; Selker et al. 2003; Martienssen and Colot 2001). This is true of most eukaryotic species, though there may be some exceptions (Simmen et al. 1999).

In humans, most repeats are the remnants of selfish DNA elements such as long interspersed nuclear elements (LINEs), short interspersed nuclear elements (SINEs), and endogenous retroviruses. These elements tend to be GC rich and the majority are heavily methylated in all tissues examined. Demethylation in somatic tissues or germ cells leads to transcriptional derepression of at least some of these elements and high levels of transcription, indicating that methylation is important for maintaining their silencing (Wa lsh et al. 1998; Bourc'his and Bestor 2004).

Inactivation and methylation of retroviruses occurs shortly after introduction into cells or embryos (Jahner et al. 1982; Stewart et al. 1982), but methylation may also be subsequent to an initial transcriptional silencing event, since changes in histone modifications can bedetectedprior to methylation ofnewlyintroducedtransgenes (Mutskov and Felsenfeld 2004). Once self-replicating DNA elements are inactivated and methylated, selective pressure will be removed and they will accumulate high rates of C'!T transitions and other mutations. Such erosion of methylation target sequences is also seen at the CpG islands of pseudogenes that arise by duplication, such as the á-globin pseudogene in human (Bird et al. 1987). In *Drosophila*, where little or no methylation is seen, silencing of self-replicating DNA such as the P element is achieved using the Polycomb/trithorax group of proteins instead, which is the major mediator of epigenetic effects in this organism.

## METHYLCYTOSINE AS AN ENDOGENOUS MUTAGEN: IMPLICATIONS IN HUMAN HEALTH

Although CpGs are relatively rare outside of the CpG islands and repeat sequences, they are not absent and can be found at low but significant levels in the promoters and coding regions of genes (Bird et al. 1985). This has important consequences for those genes that do contain them, since they are subject to high levels of transition mutations due to methylation. The effects due to deamination of methylcytosine that result in a change in sequence are, of course, distinct from those due to the effects of methylcytosine on promoter activity, which do not result in sequence changes and are therefore epigenetic. The latter effects include the methylation of trinucleotide repeats in fragile X syndrome (El-Osta 2002), the aberrant methylation and silencing of tumor suppressor genes in cancer, and the incorrect methylation of imprinted genes in certain inherited disease syndromes (see reviews cited above) and are not dealt with here.

### Inherited Disorders

Approximately 23% of all germ-line mutations responsible for genetic diseases occur at CpG positions and 90% of these are C'!T or G'!A transitions, suggesting they are due to cytosine methylation (Krawczak et al. 1998). CpG positions are affected in 40% of all point mutations on the X-linked factor VIII (*F8*) gene involved in hemophilia (Pattinson et al. 1990), while for the auto-somal *FGFR3* gene, mutation at a single CpG at codon 398 is the cause of 95% of all achondroplasia (Bellus et al. 1995; Rousseau et al. 1994). Interestingly, the frequency of mutations at CpG sites appears to be far higher in males; but a careful study by El-Maarri and colleagues has shown that this is more likely due to the higher number of replications undergone by male germ cells than anydifference in methylation, sincemature gametesof both sexes were equally methylated at non-CpG island sites, as expected (El-Maarri et al. 1998).

Two forms of mucopolysaccharidosis (MPS), types II and VII,

provide an interesting contrast in terms of methylation and mutability. MPSisa lysosomal storage disease where the inability to break down bulky glycosaminoglycan (GAG) molecules causes them to build up in the lysosomes of various organs, with detrimental effects. In both diseases, transition mutations at CpG sites are the most frequent mutations seen, accounting for 35% of point mutations at theiduronate-2-sulfatase(*IDS*)geneinvolved inMPS type II, otherwise known as Hunter syndrome (Tomatsu et al. 2004), and for 52% of point mutations at the â-glucuronidase (*GUSB*) gene involved in MPS type VII, also known as Sly syndrome (Tomatsu et al. 2002). However, at the *IDS* gene there was no correlation between the methylation status of the CpG assayed and its mutability, whereas at the *GUSB* gene a clear correlation exists between the methylation state of the CpG assayed and the number of transition mutations observed at this site. The difference may be due to the chromosomal location of the genes involved: *GUSB* is autosomal whereas *IDS* is located on the X chromosome. For the latter, it is possible that methylation of a CpG on the inactive X in the previous generation may be seen as a mutation on the active X in the next (Ohlsson et al. 2001). The same is true of the X-linked *F8* gene above, where there was a lack of correlation between methylation at a site and mutability.

An apparent difference in mutation rates among CpGs was also found in a studyof two related skin disorders. Epidermolysis bullosa simplex (EBS) and epidermolytic hyperkeratosis (EH) are related syndromes with a particularly severe phenotype where patients, usually infants, are hypersensitive to skin trauma, resulting in severe blistering (commonly called "scalded skin syndrome"). In both syndromes, the keratin proteins produced by the epidermal cells and which provide mechanical strength to the skin are faulty, leading to catastrophic collapse upon stressing. Although four keratin proteins are involved (types II K5 and I K14 in EBS and types II K1 and I K10 in EH), 6/11 of the severe cases result from mutations at a single conserved arginine residue (R125 in K14) present in the rod domain of all four proteins in an identical position (Letai et al. 1993).

This arginine is encoded by a CGC codon, where transitions result in Arg-Cys or Arg-His mutations; however, 7 other CpG-containing codons are present in the genes that could result in amino acid substitutions in highly conserved regions of the proteins. Although they did not examine methylation levels of the CpG at R125 versus the other CpGs, by generating the equivalent mutations in these latter codons, Letai et al. were able to show that none of these resulted in collapse of the keratin network and so would not be recovered in patients suffering from either syndrome. They concluded that hypermutability of the CpG at R125 is due to a combination of the high rate of transition at this site and its crucial location in the protein. The apparent difference in mutation rates among CpGs here may therefore really be due to mutations at the other sites not resulting in a visible phenotype.

## Cancer

CpG mutations in the germline can lead to inherited disease, as we have seen, butitisalsotruethat CpGmutations in somatic tissues can lead toinactivation of tumor suppressor genes and cancer. Perhaps the best-known case is the p53 (*TP53*) gene in humans, inactivating mutations of which are found in half of all human tumors, making it the most common genetic alteration found in cancer (Hollstein et al. 1991). The majority of mutations at this gene are mis-sense mutations, and according to the R9 release of the IARC *TP53* mutation database, 49% of 264 germline mutations and 24% of a total 19,809 somatic mutations are G:C'!A:T transitions at CpGs, making this the most frequent type of mutation seen in both the somatic and germline categories overall.

Of the 42 CpGs in the *TP53* gene, three of these—at codons 175, 248, and 273—account for 19% of all mutations and are considered "hotspots" (Ory et al. 1994). These observations, coupled to the fact that all of the CpGs in *TP53* are methylated in all tissues examined (Tornaletti and Pfeifer 1995), could be interpreted to mean that deamination of methylcytosine and poor repair is the major cause of mutation here.

Careful analysis of the data suggests, however, that some of the mutation bias towards the CpG dinucleotide is not due to failure to repair T/G mismatches. The CpG dinucleotides, and in particular those at codons 158, 248, and 273, are hotspots for *TP53* mutation in lung cancer, but in this case it is not a transition but a G:C'!T:A transversion instead. In fact, G'!T transversions at these CpG-containing codons of *p53* are very common in lung, head, and neck cancers, where they are associated with cigarette smoking (Soussi and Beroud 2003). BPDE, the ultimate carcinogenic metabolite arising from cigarette smoke, forms covalent chemical adducts on the N2 position of guanine.

In two elegant papers, Denissenko and colleagues showed that BPDE formed adducts preferentially at codons 157, 248, and 273 of *TP53*, but only when the DNA is methylated; when unmethylated, the site preference was far less marked, suggesting that methylation of the neighboring cytosine was promoting adduct formation (Denissenko et al. 1996, 1997) and the same is true of other smoke carcinogens (Pfeifer et al. 2002). (While all the major adduct sites are at or near CpG dinucleotides, it is not the case that all CpG sites are hotspots for adduct formation, however.) UV sunlight (Tommasi et al. 1997) and mitomycin C also have a preferential affinity for the CpG dinu-cleotide (Millard and Beachy 1993), implyingthat it is not only an endogenous promutagenic factor but also a target for several exogenous carcinogens.

Even for the G:C'!T:A transition events at CpG dinucleotides in *TP53*, not all of these events can be interpreted as causing a decrease or loss of protein function in tumors. A small fraction of these changes may be neutral in terms of selection (Soussi and Beroud 2003). We can estimate what this fraction is by examination of the database. A target cytosine occurs on both strands of the DNA in a CpG dinucleotide and methylation, deamination, and repair of both cytosines might be expected to occur at similar rates.

This is measurable, given a large enough data set such as that for *TP53*, since a failure to repair the coding strand will lead to a C'!T transition at the first base, while failure to repair the non-coding strand will lead to a G'!A transition at the second base. Examination of codons 248 and 273 bear out the idea of equal rates of reaction at each site, since the number of C'!T transitions equals the number of G'!A transitions (Soussi and Beroud 2003). Transitions in both positions lead to amino acid changes and these changes affect p53 function when engineered in vitro (Ory et al. 1994). At codon 175, however, there is a marked inequality in mutation rate, with G'!A transitions far outweighing those involving C'!T.

Whileboth types of transition at codon 175 would again lead to changes in residue (Arg'!His and Arg'!Cys, respectively), the former change completely impairs protein function in vitro (Ory et al. 1994) and is associated with very poor prognosis in colorectal cancer (Goh et al. 1995) while the rarer C'!T transition seems to have no effect on function (Ory et al. 1994). A similar situation exists for another CpG transition hotspot, codon 282 (Soussi and Beroud 2003). Alterations in the genes that give a growth advantage to the tumor are expected to be over-represented in the database and this seems to be the situation for codons 175 and 282, where one of the possible transitions at the CpGs involved is far more common than the other. The presence of the less-common mutation at these sites in some tumors is likely, in t his case, to represent co-selection for a second mutation elsewhere in the gene, so values for rates of inactivating mutation at CpG sites will therefore be artificially inflated by the presence of this small number of effectively neutral mutations (Soussi and Beroud 2003).

Notwithstanding these effects due to exogenous carcinogen preference for CpGsand the small number of co-selectedmutations, it is clear that most point mutations at *TP53* occur at CpG dinucleotides due to endogenous mutage-nesis, i.e., deamination. In other tumor suppressors too, nucleotide changes consistent with deamination and failure of repair are seen at CpG sequences. An extensive database of mutations also exists for the retinoblastoma (*RB1*) gene and shows that transition mutations at 12 of the 15 CGA codons in the open reading frame (ORF) account for 76% of the nonsense mutations seen and are by far the most prevalent type of mutation at this gene.

That this is probably a result of methylation is supported by the finding that most of these sites in the ORF are methylated and that the unmethylated CpG island at the promoter shows no suchtransition mutations in tumors (Mancini et al. 1997). Data on other genes mutated in cancer bear out the general trends seen at the better-characterized *TP53* and *RB1* loci, with hypermutability of CpGs, often located at arginine codons, resulting in hotspots for point mutations in genes such as *GNAS1* (aka the *gsp* oncogene) in pituitary tumors (Lania et al. 2003; Landis et al. 1989), *PTEN* in endometrial carcinomas and glioblastomas (Bonneau and Longy 2000), *AR* in prostate cancer (Gottlieb et al. 1997), and many others. A direct role for a G/T mismatch-specific repair enzyme in cancer has also recently been demonstrated in mice, as we shall see below.

## REPAIR OF METHYLCYTOSINE DEAMINATION BY GLYCOSYLASES IN MAMMALS

### Discovery of G/T Mismatch-Specific Repair in Eukaryotes

In contrast to mismatches generated during DNA replication, which are corrected in favor of the parental strand as discussed above, G/T mispairs arising from 5meC deamination in the resting DNA must be processed to restore the original cytosine base. The existence of a G/T mismatch-specific repair pathway in eukaryotes was first demonstrated in an African green monkey cell line (CV-1) by Brown and Jiricny (1987). Synthetic DNA containing a G/T mismatch was inserted into the genome of Simian virus (SV)40 and transfected into CV-1 cells. Analysis of recovered viral DNA revealed that mismatches were efficiently repaired and over 90% corrected to G/C pairs, i.e., in favor of guanine.

The biased repair also occurred for mismatches placed in a sequence context other than CpG, suggesting the presence of a common repair pathway that recognizes and acts on the mismatched base itself. The enzymatic activity catalyzing the removal of thymine was subsequently detected in nuclear extract from HeLa cells using a synthetic G/T mismatch-containing heterodu-plex as substrate (Wiebauer and Jiricny 1989). Further work from the same lab led to the characterization of the first thymine-specific glycosylase (TDG) (Wiebauer and Jiricny 1990), purification of the enzyme (Neddermann, Jiricny 1993), and the cloning of the *TDG* gene (Neddermann et al. 1996).

In the VSP pathway in bacteria, removal of mismatched T depends on a specific endonuclease Vsr, which cleaves the phosphodiester bond 5' to the mismatched thymine to trigger strand-specific, exonucleolytic degradation and re-synthesis of a short stretch of DNA strand. In contrast, the initiation of G/T mismatch repair in mammals relies on the cleavage of the glycosylic bond between the thymine base and the ribose,

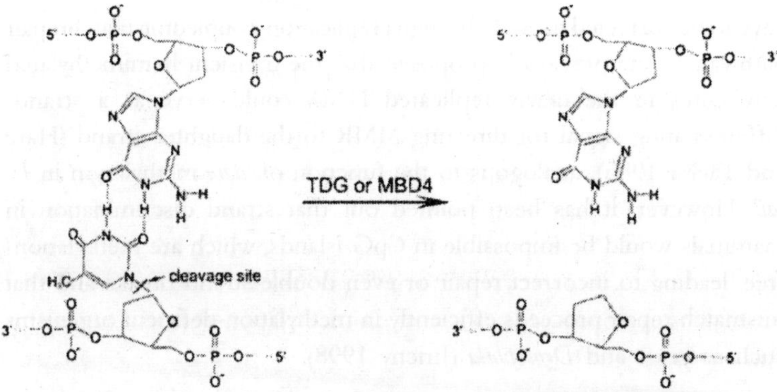


***Fig. Excision of mismatched thymine by TDG and MBD4. A G/T mispair in DNA is recognized by a glycosylase, TDG or MBD4, and the thymine base is excised by hydrolytic cleavage of the N-glycosylic bond, creating an abasic site***

creating an abasic site opposite the guanine. The abasic site serves as a secondary signal to start the downstream events of the BER pathway that are common for the repair of a variety of damaged bases (reviewed by Dianovet al. 2003; Lindahl 2001). In brief, the presence of apurinic/AP sites is sensed by an AP endonuclease (APE) that incises the affected strand 5' of the remaining phosphodeoxyribose residue. Through the concerted actions of APE endonuclease and DNA polymerase p, exonucleolytic degradation and re-synthesis proceed in a region spanning several nucleotides. DNA ligase III completes the BER pathway by sealing the nick on the repaired strand. The action of thymine glycosylase-mediated BER is subject to time constraints. As a natural base, the thymine, if not repaired while mispaired with G, will escape correction when DNA replication takes place. It is possible that coordination with cell-cycle progression exists to ensure the efficiency of G/T- specific BER.

## Does Methylation Play a Role in Directing Replication-Coupled Mismatch Repair?

Base mismatches, including G/T mispairs, arising from erroneous incorporation of a nucleotide during DNA replication should always be repaired in favor of the sequence of the parental strand. For instance, G/T mismatches due to misincorporation of G in the daughter strand have to be corrected to A/T througha replication-coupledmismatchrepair pathway. It was previously proposed that the transient hemimethylated CpG sites in the newly replicated DNA could serve as a strand-differentiating signal for directing MMR to the daughter strand (Hare and Taylor 1985), analogous to the function of *dam* methylation in *E. coli.* However, it has been pointed out that strand discrimination in mammals would be impossible in CpG islands, which are methylation-free, leading to incorrect repair or even double-strand breaks and that mismatch repair proceeds efficiently in methylation-deficient organisms such as yeast and *Drosophila* (Jiricny 1998).

Later experiments also suggested that methylation did not in fact direct repair in vitro (Drummond and Bellacosa 2001). Although it is still unclear how strand discrimination occurs in eukaryotes, it is thought that it may occur at the replication complex itself (Jiricny 1998). Methylation may still somehow be involved in those eukaryotes that have it, as the maintenance methyltransferase DNMT1, which is also associated with replication foci at S phase, has recently been implicated in mismatch repair (Guo et al. 2004; Wang and James Shen 2004). In addition, the mismatch repair protein MLH1 interacts with MBD4, a methyl CpG binding protein and glycosylase (Bellacosa et al. 1999; Parsons 2003). This circumstantial evidence suggests the existence of cross-talk between MMR components and methylation signals, though the precise roles of DNMT1 and MBD4 remain to be defined.

## Excision of Deaminated Methylcytosines by TDG and MBD4

Among the eight DNA glycosylases found in the human genome, TDG and the more recently discovered MBD4 are the only two enzymes able to correct G/T mismatches to G/C (Wood et al. 2001). These two enzymes are thought to play a central role in the detection and excision of the mismatched thymine to initiate the BER process. The biochemistry and biology of TDG have been covered by three excellent reviews (Hardeland et al. 2001; Schärer and Jiricny 2001; Waters and Swann 2000) and will not be dealt with in any detail here.

The inefficiency of repair by TDG may partly explain the high rate of mutation at CpG sites in the human genome already noted in Sect. 2 above. The enzyme is limited in two ways. First, it has a very low $K_{cat}$, even on its preferred target, CpG (0.91 min-[1]), compared with a $K_{cat}$ of 2,500 min-[1] for UDG: In other words, UDG could process more than 2,000 mismatches while TDG was still struggling with its first (Waters and Swann 1998). Second, it exhibits product inhibition: Upon excision of the mismatched base in vitro, TDG remains bound with DNA at the abasic site with high affinity, resulting in an extremely low enzymatic turnover on the order of 5-10 h. However, this latter rate-limiting step might be regulated in vivo by sumoylation of TDG (Hardeland et al. 2002) and its interaction with XPC, a protein involved in nu-cleotide excision repair (Shimizu et al. 2003), either of which has a stimulating effect on the release of TDG from the abasic site.

Binding of the second BER component APE also releases TDG, effectively coupling the first and second steps of the repair pathway in this fashion (Waters et al. 1999). This is also a point of control: Acetylation of TDG by the transcriptional coactivator CBP interferes with this interaction and presumably with the displacement of TDG by APE from the abasic site, thus exerting an inhibitory role (Tini et al. 2002). TDG is also proficient in vitro in the removal of other deamination products derived from cytosines, including uracils (Krokan et al. 2002; Hardeland et al. 2003), and is implicated in transcriptional activation (Tini et al. 2002).

Given the multifunctional potential of TDG, carefully controlled experiments using cultured cells and/or animal models need to be carried out to verify its long-proposed function in the elimination of methylcytosine deamination products in vivo.

MBD4, the other mammalian enzyme that can repair G/T mismatches, was independently discovered as a member of the methyl-CpG DNA-binding protein family (Hendrich and Bird 1998) and as an interacting partner of mismatch repair protein MLH1 (Bellacosa et al. 1999). In addition to a C-terminal glycosylase domain unrelated to TDG, it contains an N-terminal MBD that is absent in TDG. Despite the phylogenetic divergence, MBD4 has similar substrate specificity and can efficiently remove T or U from a mismatch as can TDG (Hendrich et al. 1999). Interestingly, the MBD domain binds DNA preferentially at the sequence containing 5' 5meCpG/ 3'TpG, a site formed due to methylcytosine deamination atmethyl-CpG dinucleotides. This DNA-binding property, in combination with the thymine glycosylase activity, makes MBD4 appear more suited for the repair of methylcytosine deamination products than TDG, although the functional coordination of the two moieties in MBD4 needs to be addressed experimentally. A direct demonstration that a high rate of C^T transitions at CpG sites can be caused by a failure in the G-T mismatch repair machinery was provided in 2002 by two groups (Millar et al. 2002; Wong et al. 2002), who used powerful transgenic approaches to address the consequences of such failure. Both groups generated a loss-of-function mutation in the *Mbd4* gene, then crossed these mice with "Big Blue" reporter mice (Kohler et al. 1991), which allow direct measurement of the rate of mutation using a unique recoverable ë transgene. A highly significant increase in G/C'!A/T transitions at CpG sites in all mutant mice was seen without affecting other mutation categories (Millar et al. 2002; Wong et al. 2002), and this correlated with high levels of methylation in vivo at all of the CpGs assessed (Millar et al. 2002).

To show that this increase in CpG mutability could lead to an increased risk of cancer, both groups crossed *Mbd4* knockout mice to others carrying mutations at the adenomatous polyposis coli (*Apc*) gene, which predispose the mice to multiple intestinal neoplasia. There was a significant reduction in survival for *Mbd4* knockout mice carrying a heterozygous mutation at *Apc* compared to wildtype littermates and a small but significant increase in tumor number (Millar et al. 2002; Wong et al. 2002). In tumor tissues, alterations at CpG sites were greatly increased in the *Mbd4*$^{-/-}$ mice compared to their *Mbd4*$^{+/+}$ sibs, suggesting that the increased mutability of the dinucleotide was at least partly responsible.

That the effects of this mutation were not even more marked is almost certainly due to the presence of the alternative G/T mismatch repair enzyme TDG in these mice; but these experiments nevertheless provide convincing evidence that methylcytosine deamination must be occurring at high rates in vivo and that MBD4 protein is crucial for its repair. Bearing out these results are the findings that mutations in the human *MBD4* gene are frequently seen in colorectal tumors showing microsatellite-instability (MSI), though these mutations occur downstream of the MMR gene mutations and therefore are not likely to be a primary cause of MSI (Riccio et al. 1999; Bader et al. 1999; Yamada et al. 2002).

Nevertheless, the exact assessment of relative contributions of the two thymine glycosylases in counteracting the mutability of methylcytosines in vivo awaits the establishment of an animal model deficient in TDG. Conceivably, the two glycosylases could have specialized and complementary roles in safeguarding against the deamination threat posed by methylcytosines in different parts of a mammalian genome. It has been suggested that MBD4 may be targeted to more transcriptionally inactive regions with high methylation density through its MBD motif, while TDG could be localized more to transcribed regions by virtue of its association with transcriptional coactivators (Hardeland et al. 2003).

## Restoration of Methylation After G/T Specific Repair

Even when the resultant C'!T transition is efficiently repaired back to cy-tosine, methylcytosine deamination could still potentially have a profound impact on genome stability and gene expression. The restored cytosine at the mismatch site and those other newly incorporated cytosines in the short patch around it replaced by the action of the BER pathway need to be remethylated prior to the onset of DNA replication. If hemimethylated CpGs in the repaired region are not converted into fully methylated sites, passive demethylation occurs after subsequent cell divisions. If DNA methylation exerted a gene regulatory function solely through effects on chromatin remodeling on a regional level, loss of methylation at a single site might have no effect. However, changes at an individual CpG site within the binding sequence of a transcriptional factor can sometimes interfere with protein–DNA interaction. CTCF, a protein with an important role in the imprinting of the *H19/Igf2* locus, binds specifically to the unmethylated target site (Hark et al. 2000) while Kaiso, a component of the N-CoR complex, depends on cytosine methylation in its binding sites to mediate transcriptional repression (Yoon et al. 2003). In such cases, demethylation at a single key site may lead to the dysregulation of gene expression, for example erroneous activation of an oncogene in tumorigenesis.

We therefore hypothesize that coordinated remethylation of cytosines may occur in conjunction with DNA repair in methylated regions of the genome. In the repair of 5meC deamination products, the replacement cytosines would be subject to a remethylation process coupled to the G/T-specific BER pathway. Among the four active DNA methyltransferases known to date, the maintenance enzyme Dnmt1 might be a favored candidate for the remethyla-tion of repaired cytosines because of its strong preference for hemimethylated sites and ubiquitous expression at all developmental stages. However, the two de novo methyltransferases Dnmt3a and Dnmt3b are also present in various somatic tissues, albeit at low levels, and could therefore also be involved in remethylation of cytosines derived from G/T mismatch repair.

Some preliminary evidence for the proposed coupling of remethylation to the G/T-specific BER pathway comes from our observation that Dnmt3a interacts with TDG (Y.Q. Li, P.Z. Zhou, G.L. Xu, unpublished data). TDG was isolated repeatedly as an interaction partner of Dnmt3a in yeast two-hybrid screening experiments, and physical interaction could be confirmed by glutathione *S*-transferase (GST) pull-down assays in vitro. Moreover, the glycosylase activity of recombinant TDG was greatly stimulated by the addition of Dnmt3a but not Dnmt3L and Dnmt2 proteins. The physical and functional interaction between TDG and Dnmt3a suggests a potential link

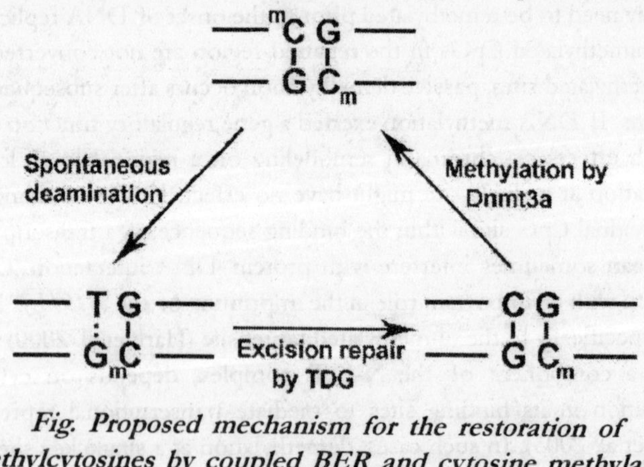


***Fig.* Proposed mechanism for the restoration of methylcytosines by coupled BER and cytosine methylation processes**

between BER and DNA methylation in the maintenance of genetic and epi-genetic stability of the mammalian genome. Studies are now underway to examine the functional coordination between G/T-specific BER and cytosine methylation in vivo in cells transfected with G/T mismatch-containing DNA. The need for remethylation of cytosines is conceivably widespread in a variety of DNA repair events involving strand re-synthesis in a methylated ge-nomic region. Failure of remethylation might be implicated in the generation of abnormal genomic methylation patterns in various diseases, including cancer. Further investigation is needed to demonstrate the biological relevance of cytosine remethylation in various physiological settings and to identify factors that ensure coupling to repair.

## Potential Role of Deamination and Glycosylases in Demethylation

Reprogramming of genomic methylation patterns involving active demethy-lation in a replication-independent manner is likely to occur at two developmental stages. In the primordial germ cells, genome-wide demethylation happens during a very short window to allow resetting of sex-specific methy-lation patterns in the gametes (Hajkova et al. 2002; Lees-Murdock et al. 2003; Li et al. 2004; Tada et al. 1997). In early embryogenesis, the paternal genome in the zygote is also subject to dramatic global demethylation before DNA replication (Mayer et al. 2000; Oswald et al. 2000). Despite several false starts, the enzyme(s) and the biochemical mechanism underlying active demethylation have not been convincingly identified yet (see Bird 2002 for a recent review).

## Demethylation via Direct Excision of Methylcytosines by Glycosylases

Active DNA demethylation could occur simply by the direct excision of a methylcytosine base by glycosylases. Both TDG and MBD4 are found to be active on methylcytosines in addition to mismatched thymines (Zhu et al. 2000a, b). However, the efficiency of 5meC excision is much lower in comparison with that of T mispaired with G. This raises the possibility that the minor in vitro 5meC glycosylase activity may have little physiological relevance. Overexpression of transfected chicken TDG led to activation and promoter demethylation of a co-transfected ecdysone-retinoic acid responsive reporter gene in 293 cells, however (Zhu et al. 2001), suggesting that the enzyme may be important for demethylation in vivo. Other studies have also shown physical and functional interactions of TDG with transcription factors (Missero et al. 2001), hormone receptors (Chen et al. 2003; Um et al. 1998), and chromatin-remodeling proteins (Tini et al. 2002).

Weighing against a general involvement of TDG in active demethylation is the fact that overexpression of the enzyme in transfected cells was not sufficient to cause generalized demethylation of the genome (Zhu et al. 2001). One group has also reported that the transcriptional repressor MBD2 has demethylating ability (Detich et al. 2002; Bhattacharya et al. 1999); however, other labs have been unable to replicate these findings (Wolffe et al. 1999; Bird 2002). Further investigation will be required to determine whether TDG, or indeed MBD2, could have any role in active demethylation.

Intriguing indirect evidence for the role of glycosylases in DNA demethy-lation in association with transcriptional activation has emerged recently in plants. An *Arabidopsis* homolog (ROS1) has a clear role in the repression of DNA methylation-mediated transgene silencing, presumably by its demethylation activity (Gong et al. 2002). Mutations in the *ROS1* gene cause transcriptionalsilencingof a transgene and the homologous endogenous gene as a consequence of promoter hypermethylation. Recombinant ROS1 protein is a bifunctional glycosylase/lyase, cleaving the phosphodiester backbone of double-stranded DNA at sites where a methylated cytosine base has been removed by the glycosylase activity.

A more recent paper revealed another plant glycosylase (DME) involved in imprinting of the MEDEA Polycomb gene in *Arabidopsis* (Xiao et al. 2003). DME activates transcription of the maternal allele of the imprinted *MEDEA* gene in the central cell of the female gametophyte, antagonizing the suppressing effect of cytosine methylation by MET1, an ortholog of the mammalian methyltransferase Dnmt1 (Xiao et al. 2003). The glycosylase activity of DME is required for this transcriptional activation, supposedly by causing demethylation of the MEDEA promoter (Choi et al. 2004). Interestingly, the two plant glycosylases ROS1 and DME and the mammalian glycosylase MBD4 are members of the same HhH family of bifunctional glycosylase/lyase enzymes represented by MutY and endonu-clease III from *E. coli*. More detailed biochemical and functional studies are needed to confirm their roles in DNA demethylation.

# Bibliography


B S Bedi: *DNA Testing*, Mohit Publication, Delhi, 2006

Bharat Singh: *Recombinant DNA and Biotechnology*, Pointer, Delhi, 2007.

Brian K. Nunnally: *Analytical Techniques in DNA Sequencing*, Taylor and Francis Group, Delhi, 2009.

Crick, F. J.: *Life Itself: Its Origin and Nature*, New York, Simon and Schuster, 1981.

Crick, F.: *Life Itself: Its Origin and Nature*, New York, W.W. Norton, 1982.

Fox, S. W.: *A model for protocellular coordination of nucleic acid and protein synthesis*, Tokyo, Japan Science Society Press, 1981.

J J Dhruve; U K Kandoliya; S V Patel and B A Golakiya: *DNA : A Bridge Between Biochemistry and Biotechnology*, New India Publishing Agency, 2008.

Kurt Weising, Hilde Nybom, Kirsten Wolff and Gunter Kahl: *DNA Fingerprinting in Plants : Principles, Methods and Applications*, Taylor and Francis Group, Delhi, 2009.

Peter Walker and Rohana Subasinghe: *DNA Based Molecular Diagnostic Techniques : Research Needs for Standardization and Validation*, Daya, Delhi, 2005.

Ronald L Phillips and Indra K Vasil: *DNA-Based Markers in Plants*, Springer, Delhi, 2007.

S K Mangal: *DNA Markers in Plant Improvement*, Gene Tech Book, Delhi, 2007.

Stebbins, G. L.: *Darwin to DNA, Molecules to Humanity*, San Francisco, W. H. Freeman, 1982.

Watson, J. D.: *The Double Helix*, New York, Antheneum, 1968.

# Index